FUNCTIONAL ANALYSIS

MONOGRAPHS AND TEXTBOOKS IN
PURE AND APPLIED MATHEMATICS

1. *K. Yano*, Integral Formulas in Riemannian Geometry (1970) *(out of print)*
2. *S. Kobayashi*, Hyperbolic Manifolds and Holomorphic Mappings (1970) *(out of print)*
3. *V. S. Vladimirov*, Equations of Mathematical Physics (A. Jeffrey, editor A. Littlewood, translator) (1970) *(out of print)*
4. *B. N. Pshenichnyi*, Necessary Conditions for an Extremum (L. Neustadt, translation editor; K. Makowski, translator) (1971)
5. *L. Narici, E. Beckenstein, and G. Bachman*, Functional Analysis and Valuation Theory (1971)
6. *S. S. Passman*, Infinite Group Rings (1971)
7. *L. Dornhoff*, Group Representation Theory (in two parts). Part A: Ordinary Representation Theory. Part B: Modular Representation Theory (1971, 1972)
8. *W. Boothby and G. L. Weiss (eds.)*, Symmetric Spaces: Short Courses Presented at Washington University (1972)
9. *Y. Matsushima*, Differentiable Manifolds (E. T. Kobayashi, translator) (1972)
10. *L. E. Ward, Jr.*, Topology: An Outline for a First Course (1972) *(out of print)*
11. *A. Babakhanian*, Cohomological Methods in Group Theory (1972)
12. *R. Gilmer*, Multiplicative Ideal Theory (1972)
13. *J. Yeh*, Stochastic Processes and the Wiener Integral (1973) *(out of print)*
14. *J. Barros-Neto*, Introduction to the Theory of Distributions (1973) *(out of print)*
15. *R. Larsen*, Functional Analysis: An Introduction (1973) *(out of print)*
16. *K. Yano and S. Ishihara*, Tangent and Cotangent Bundles: Differential Geometry (1973) *(out of print)*
17. *C. Procesi*, Rings with Polynomial Identities (1973)
18. *R. Hermann*, Geometry, Physics, and Systems (1973)
19. *N. R. Wallach*, Harmonic Analysis on Homogeneous Spaces (1973) *(out of print)*
20. *J. Dieudonné*, Introduction to the Theory of Formal Groups (1973)
21. *I. Vaisman*, Cohomology and Differential Forms (1973)
22. *B. -Y. Chen*, Geometry of Submanifolds (1973)
23. *M. Marcus*, Finite Dimensional Multilinear Algebra (in two parts) (1973, 1975)
24. *R. Larsen*, Banach Algebras: An Introduction (1973)
25. *R. O. Kujala and A. L. Vitter (eds.)*, Value Distribution Theory: Part A; Part B: Deficit and Bezout Estimates by Wilhelm Stoll (1973)
26. *K. B. Stolarsky*, Algebraic Numbers and Diophantine Approximation (1974)
27. *A. R. Magid*, The Separable Galois Theory of Commutative Rings (1974)
28. *B. R. McDonald*, Finite Rings with Identity (1974)
29. *J. Satake*, Linear Algebra (S. Koh, T. A. Akiba, and S. Ihara, translators) (1975)

30. *J. S. Golan*, Localization of Noncommutative Rings (1975)
31. *G. Klambauer*, Mathematical Analysis (1975)
32. *M. K. Agoston*, Algebraic Topology: A First Course (1976)
33. *K. R. Goodearl*, Ring Theory: Nonsingular Rings and Modules (1976)
34. *L. E. Mansfield*, Linear Algebra with Geometric Applications: Selected Topics (1976)
35. *N. J. Pullman*, Matrix Theory and Its Applications (1976)
36. *B. R. McDonald*, Geometric Algebra Over Local Rings (1976)
37. *C. W. Groetsch*, Generalized Inverses of Linear Operators: Representation and Approximation (1977)
38. *J. E. Kuczkowski and J. L. Gersting*, Abstract Algebra: A First Look (1977)
39. *C. O. Christenson and W. L. Voxman*, Aspects of Topology (1977)
40. *M. Nagata*, Field Theory (1977)
41. *R. L. Long*, Algebraic Number Theory (1977)
42. *W. F. Pfeffer*, Integrals and Measures (1977)
43. *R. L. Wheeden and A. Zygmund*, Measure and Integral: An Introduction to Real Analysis (1977)
44. *J. H. Curtiss*, Introduction to Functions of a Complex Variable (1978)
45. *K. Hrbacek and T. Jech*, Introduction to Set Theory (1978)
46. *W. S. Massey*, Homology and Cohomology Theory (1978)
47. *M. Marcus*, Introduction to Modern Algebra (1978)
48. *E. C. Young*, Vector and Tensor Analysis (1978)
49. *S. B. Nadler, Jr.*, Hyperspaces of Sets (1978)
50. *S. K. Segal*, Topics in Group Rings (1978)
51. *A. C. M. van Rooij*, Non-Archimedean Functional Analysis (1978)
54. *L. Corwin and R. Szczarba*, Calculus in Vector Spaces (1979)
53. *C. Sadosky*, Interpolation of Operators and Singular Integrals: An Introduction to Harmonic Analysis (1979)
54. *J. Cronin*, Differential Equations: Introduction and Quantitative Theory (1980)
55. *C. W. Groetsch*, Elements of Applicable Functional Analysis (1980)
56. *I. Vaisman*, Foundations of Three-Dimensional Euclidean Geometry (1980)
57. *H. I. Freedman*, Deterministic Mathematical Models in Population Ecology (1980)
58. *S. B. Chae*, Lebesgue Integration (1980)
59. *C. S. Rees, S. M. Shah, and C. V. Stanojević*, Theory and Applications of Fourier Analysis (1981)
60. *L. Nachbin*, Introduction to Functional Analysis: Banach Spaces and Differential Calculus (R. M. Aron, translator) (1981)
61. *G. Orzech and M. Orzech*, Plane Algebraic Curves: An Introduction Via Valuations (1981)
62. *R. Johnsonbaugh and W. E. Pfaffenberger*, Foundations of Mathematical Analysis (1981)
63. *W. L. Voxman and R. H. Goetschel*, Advanced Calculus: An Introduction to Modern Analysis (1981)
64. *L. J. Corwin and R. H. Szcarba*, Multivariable Calculus (1982)
65. *V. I. Istrătescu*, Introduction to Linear Operator Theory (1981)
66. *R. D. Järvinen*, Finite and Infinite Dimensional Linear Spaces: A Comparative Study in Algebraic and Analytic Settings (1981)

67. *J. K. Beem and P. E. Ehrlich,* Global Lorentzian Geometry (1981)
68. *D. L. Armacost,* The Structure of Locally Compact Abelian Groups (1981)
69. *J. W. Brewer and M. K. Smith, eds.,* Emmy Noether: A Tribute to Her Life and Work (1981)
70. *K. H. Kim,* Boolean Matrix Theory and Applications (1982)
71. *T. W. Wieting,* The Mathematical Theory of Chromatic Plane Ornaments (1982)
72. *D. B. Gauld,* Differential Topology: An Introduction (1982)
73. *R. L. Faber,* Foundations of Euclidean and Non-Euclidean Geometry (1983)
74. *M. Carmeli,* Statistical Theory and Random Matrices (1983)
75. *J. H. Carruth, J. A. Hildebrant, and R. J. Koch,* The Theory of Topological Semigroups (1983)
76. *R. L. Faber,* Differential Geometry and Relativity Theory: An Introduction (1983)
77. *S. Barnett,* Polynomials and Linear Control Systems (1983)
78. *G. Karpilovsky,* Commutative Group Algebras (1983)
79. *F. Van Oystaeyen and A. Verschoren,* Relative Invariants of Rings: The Commutative Theory (1983)
80. *I. Vaisman,* A First Course in Differential Geometry (1984)
81. *G. W. Swan,* Applications of Optimal Control Theory in Biomedicine (1984)
82. *T. Petrie and J. D. Randall,* Transformation Groups on Manifolds (1984)
83. *K. Goebel and S. Reich,* Uniform Convexity, Hyperbolic Geometry, and Nonexpansive Mappings (1984)
84. *T. Albu and C. Năstăsescu,* Relative Finiteness in Module Theory (1984)
85. *K. Hrbacek and T. Jech,* Introduction to Set Theory, Second Edition, Revised and Expanded (1984)
86. *F. Van Oystaeyen and A. Verschoren,* Relative Invariants of Rings: The Noncommutative Theory (1984)
87. *B. R. McDonald,* Linear Algebra Over Commutative Rings (1984)
88. *M. Namba,* Geometry of Projective Algebraic Curves (1984)
89. *G. F. Webb,* Theory of Nonlinear Age-Dependent Population Dynamics (1985)
90. *M. R. Bremner, R. V. Moody, and J. Patera,* Tables of Dominant Weight Multiplicities for Representations of Simple Lie Algebras (1985)
91. *A. E. Fekete,* Real Linear Algebra (1985)
92. *S. B. Chae,* Holomorphy and Calculus in Normed Spaces (1985)
93. *A. J. Jerri,* Introduction to Integral Equations with Applications (1985)
94. *G. Karpilovsky,* Projective Representations of Finite Groups (1985)
95. *L. Narici and E. Beckenstein,* Topological Vector Spaces (1985)
96. *J. Weeks,* The Shape of Space: How to Visualize Surfaces and Three-Dimensional Manifolds (1985)
97. *P. R. Gribik and K. O. Kortanek,* Extremal Methods of Operations Research (1985)
98. *J.-A. Chao and W. A. Woyczynski, eds.,* Probability Theory and Harmonic Analysis (1986)
99. *G. D. Crown, M. H. Fenrick, and R. J. Valenza,* Abstract Algebra (1986)
100. *J. H. Carruth, J. A. Hildebrant, and R. J. Koch,* The Theory of Topological Semigroups, Volume 2 (1986)

101. *R. S. Doran and V. A. Belfi*, Characterizations of C*-Algebras: The Gelfand-Naimark Theorems (1986)

102. *M. W. Jeter*, Mathematical Programming: An Introduction to Optimization (1986)

103. *M. Altman*, A Unified Theory of Nonlinear Operator and Evolution Equations with Applications: A New Approach to Nonlinear Partial Differential Equations (1986)

104. *A. Verschoren*, Relative Invariants of Sheaves (1987)

105. *R. A. Usmani*, Applied Linear Algebra (1987)

106. *P. Blass and J. Lang*, Zariski Surfaces and Differential Equations in Characteristic p $>$ 0 (1987)

107. *J. A. Reneke, R. E. Fennell, and R. B. Minton*. Structured Hereditary Systems (1987)

108. *H. Busemann and B. B. Phadke*, Spaces with Distinguished Geodesics (1987)

109. *R. Harte*, Invertibility and Singularity for Bounded Linear Operators (1988).

110. *G. S. Ladde, V. Lakshmikantham, and B. G. Zhang*, Oscillation Theory of Differential Equations with Deviating Arguments (1987)

111. *L. Dudkin, I. Rabinovich, and I. Vakhutinsky*, Iterative Aggregation Theory: Mathematical Methods of Coordinating Detailed and Aggregate Problems in Large Control Systems (1987)

112. *T. Okubo*, Differential Geometry (1987)

113. *D. L. Stancl and M. L. Stancl*, Real Analysis with Point-Set Topology (1987)

114. *T. C. Gard*, Introduction to Stochastic Differential Equations (1988)

115. *S. S. Abhyankar*, Enumerative Combinatorics of Young Tableaux (1988)

116. *H. Strade and R. Farnsteiner*, Modular Lie Algebras and Their Representations (1988)

117. *J. A. Huckaba*, Commutative Rings with Zero Divisors (1988)

118. *W. D. Wallis*, Combinatorial Designs (1988)

119. *W. Więsław*, Topological Fields (1988)

120. *G. Karpilovsky*, Field Theory: Classical Foundations and Multiplicative Groups (1988)

121. *S. Caenepeel and F. Van Oystaeyen*, Brauer Groups and the Cohomology of Graded Rings (1989)

122. *W. Kozlowski*, Modular Function Spaces (1988)

123. *E. Lowen-Colebunders*, Function Classes of Cauchy Continuous Maps (1989)

124. *M. Pavel*, Fundamentals of Pattern Recognition (1989)

125. *V. Lakshmikantham, S. Leela, and A. A. Martynyuk*, Stability Analysis of Nonlinear Systems (1989)

126. *R. Sivaramakrishnan*, The Classical Theory of Arithmetic Functions (1989)

127. *N. A. Watson*, Parabolic Equations on an Infinite Strip (1989)

128. *K. J. Hastings*, Introduction to the Mathematics of Operations Research (1989)

129. *B. Fine*, Algebraic Theory of the Bianchi Groups (1989)

130. *D. N. Dikranjan, I. R. Prodanov, and L. N. Stoyanov*, Topological Groups: Characters, Dualities, and Minimal Group Topologies (1989)

131. *J. C. Morgan II,* Point Set Theory (1990)
132. *P. Biler and A. Witkowski,* Problems in Mathematical Analysis (1990)
133. *H. J. Sussmann,* Nonlinear Controllability and Optimal Control (1990)
134. *J.-P. Florens, M. Mouchart, and J. M. Rolin,* Elements of Bayesian Statistics (1990)
135. *N. Shell,* Topological Fields and Near Valuations (1990)
136. *B. F. Doolin and C. F. Martin,* Introduction to Differential Geometry for Engineers (1990)
137. *S. S. Holland, Jr.,* Applied Analysis by the Hilbert Space Method (1990)
138. *J. Okniński,* Semigroup Algebras (1990)
139. *K. Zhu,* Operator Theory in Function Spaces (1990)
140. *G. B. Price,* An Introduction to Multicomplex Spaces and Functions (1991)
141. *R. B. Darst,* Introduction to Linear Programming: Applications and Extensions (1991)
142. *P. L. Sachdev,* Nonlinear Ordinary Differential Equations and Their Applications (1991)
143. *T. Husain,* Orthogonal Schauder Bases (1991)
144. *J. Foran,* Fundamentals of Real Analysis (1991)
145. *W. C. Brown,* Matrices and Vector Spaces (1991)
146. *M. M. Rao and Z. D. Ren,* Theory of Orlicz Spaces (1991)
147. *J. S. Golan and T. Head,* Modules and the Structures of Rings: A Primer (1991)
148. *C. Small,* Arithmetic of Finite Fields (1991)
149. *K. Yang,* Complex Algebraic Geometry: An Introduction to Curves and Surfaces (1991)
150. *D. G. Hoffman, D. A. Leonard, C. C. Lindner, K. T. Phelps, C. A. Rodger, and J. R. Wall,* Coding Theory: The Essentials (1991)
151. *M. O. González,* Classical Complex Analysis (1992)
152. *M. O. González,* Complex Analysis: Selected Topics (1992)
153. *L. W. Baggett,* Functional Analysis: A Primer (1992)

Other Volumes in Preparation

FUNCTIONAL ANALYSIS

A Primer

Lawrence W. Baggett

University of Colorado at Boulder
Boulder, Colorado

Marcel Dekker, Inc.　　　New York • Basel • Hong Kong

Library of Congress Cataloging--in--Publication Data

Baggett, Lawrence W.
  Functional analysis: a primer/Lawrence W. Baggett.
    p.   cm. -- -- (Monographs and textbooks in pure and applied
  mathematics)
  Includes bibliographical references and index.
  ISBN 0-8247-8598-3 (alk. paper)
  1. Functional analysis. I. Title. II. Series.
QA320.B26   1992
515'.7-- --dc20   91-25601
                                                                  CIP

This book is printed on acid-free paper.

MARCEL DEKKER, INC.
270 Madison Avenue, New York, New York 10016

Current printing (last digit):
10 9 8 7 6 5 4 3 2 1

PRINTED IN THE UNITED STATES OF AMERICA

PREFACE

The marriage of algebra and topology has produced many beautiful and intricate subjects in mathematics, of which perhaps the broadest is functional analysis. My aim has been to write a textbook with which graduate students can master at least some of the powerful tools of this subject. Because I think that one learns best by doing, I believe that it is critical that the students using this book in a course work the exercises. As an integral part of the book, they have been designed to provide practice in mimicking the techniques that are presented here in the proofs, as well as to lead the novice through fairly elaborate arguments that establish important additional results. The instructor is encouraged and expected to add theorems and examples from his or her own experiences and preferences, for I have quite deliberately restricted this presentation according to my own. My style is to state relatively few theorems, each having a fairly substantial proof, rather than to present a long series of lemmas. The student should read these substantial proofs with pencil in hand, making sure how each step follows from the previous ones and filling in any details that have been left to the reader.

I propose this text for a one-year course. The first six chapters constitute a general study of topological vector spaces, Banach spaces, duality, convexity, etc., concluding with a chapter that contains a number of applications to classical analysis, e.g., convolution, Green's functions, the Fourier transform, and the Hilbert transform.

I assume that the students studying from this book have completed a course in general measure theory, so that terms such as outer measure, σ-algebra, measurability, L^p spaces (including the Riesz representation theorem for $(L^p)^*$), product measures, etc. should be familiar. In addition, I freely use concepts such as separability and completeness from metric space theory (making particular use of the Baire category theorem at several points in Chapter IV), and I employ the general Stone-Weierstrass theorem on several occasions. I also think that many aspects of general topology were in fact invented to support the concepts in functional analysis, and I draw on these results in some rather deep ways. Thinking that those aspects of general topology that are most critical to this subject, e.g., product topologies, weak topologies, convergence of nets, etc., may not be covered in sufficient detail in many elementary topology courses, I go to some effort to explain these notions carefully throughout the text.

I do not intend to include here the most general cases of theorems and definitions, believing that my versions are both hard enough and deep enough for a student's first go at this subject. For example, I consider only locally compact topological spaces that are second countable, measures that are σ-finite, and Hilbert spaces that are separable. Chapter 0 is a kind of catalog for the basic results from linear algebra and topology that will be assumed.

The second half of the book centers on the Spectral Theorem in Hilbert space, the most important theorem of functional analysis in my view. Students with some elementary knowledge of Banach space theory and the Riesz Representation Theorem for $(C(X))^*$ can in fact begin with Chapter VIII, referring to the earlier chapters on those few occasions when more delicate results from locally convex analysis and dual topologies are required. I introduce early on the notion of projection-valued measures and spend some time studying operators that can be expressed as integrals against such a measure. I present the Gelfand approach to the Spectral Theorem for a bounded normal operator, for my sense is that the beauty of that approach is so spectacular that it should be experienced by every analyst, hard or soft. In Chapter XI I spend some time studying the standard classes of operators ordinarily encountered in analysis: compact, Hilbert-Schmidt, trace class, and unbounded selfadjoint. I include only a few of the large number of examples from differential and integral equations that spawned these classes of operators, leaving this addition to the instructor's choice. Indeed, my choice has been to present an introduction to the connection between operator theory and the foundations of quantum mechanics. Thus I devote Chapter VII to a brief presentation of a set of axioms for a mathematical model of experimental science. These axioms are a minor perturbation of those first introduced by G. W. Mackey for Quantum Mechanics, and my aim is to motivate the notions of projection-valued measures and unbounded selfadjoint operators by using them as models of the question-valued measures and observables of this set of axioms. However, this chapter and all references to it can be omitted without any effect on the rest of the material.

Finally, Chapter XII is devoted primarily to a development of the Implicit Function Theorem in infinite dimensions. My experience is that most beginning graduate students can well use another trek through the ideas surrounding this theorem, and what is presented here also provides a basic introduction to nonlinear functional analysis.

The bibliography includes two texts on measure theory and real analysis for reference purposes, several of the standard volumes on functional analysis for different points of view, and a number of books the interested student should consider reading after finishing this one.

It has been said that many functional analysis books are too big. They are encyclopedic; they have everything in them. Having tried to study from them, a student often leaves the course more baffled by the tool box than by the tasks to be solved. I expect students of this small textbook to be masters of the most powerful and commonly used tools.

Three classes of functional analysis students have helped my class notes evolve into a book. I valued and welcomed their compliments, their complaints, and their true partnership in developing this material. I thank them all. Many of my faculty colleagues have provided me with alternate proofs, interesting examples, and novel exercises, all of which have enriched the book, and I thank them all, too. Special thanks go to my daughter Molly for her diligent help with the indexing. Finally, I thank my wife Christy for her compliments, her complaints, and her true partnership in developing this material. More particularly, I am extremely grateful for her considerable editorial expertise, which has been of continuous help. I count myself extremely lucky to have had such encouragement and support.

As a blind person, I am incredibly indebted to many kinds of electronic and software products, for without them I could not in all probability have prepared this text at all. Confessing that I am neglecting to mention many such items, I especially thank Professor Donald Knuth for his program TEXand the American Mathematical Society for its macro package $\mathcal{AMS}$-TEX. Also I am in daily debt to the ECHO and ACCENT speech synthesizers as well as to the two computer programs PROTERM (MicroTalk) and JAWS (Henter-Joyce).

Lawrence W. Baggett

CONTENTS

PREFACE iii

0. PRELIMINARIES 1
   Hausdorff Maximality Principle
   Cartesian Products
   Vector Spaces
   Linear Transformations
   Kernel and Range
   Basic Topological Definitions
   Weak Topologies
   Product Spaces
   Quotient Spaces

I. THE RIESZ REPRESENTATION THEOREM 11
   Vector Lattices and Stone's Axiom
   A Fundamental Representation Theorem
   Second Countable Locally Compact Spaces
   Riesz Representation Theorems
   Complex Riesz Representation Theorem

II. THE HAHN-BANACH EXTENSION THEOREMS
    AND EXISTENCE OF LINEAR FUNCTIONALS 27
    Hahn-Banach Theorem, Positive Cone Version
    Seminorms and Subadditive Functionals
    Hahn-Banach Theorem, Seminorm Version
    Nets in Topological Spaces
    Existence of Invariant Linear Functionals
    Banach Limits
    Finitely-Additive Translation-Invariant Measures
    Hahn-Banach Theorem, Complex Version

III. TOPOLOGICAL VECTOR SPACES AND
     CONTINUOUS LINEAR FUNCTIONALS 43
     Axioms for a Topological Vector Space
     Subspaces and Quotient Spaces
     Continuous Linear Transformations
     Weak Vector Space Topologies
     Schwartz Space $\mathcal{S}$
     Locally Convex Topologies
     Existence of Continuous Seminorms
     Hahn-Banach Theorem for Locally Convex Spaces
     Separation Theorem
     Extreme Points
     Krain-Milman Theorem
     Choquet's Theorem

IV. NORMED LINEAR SPACES
    AND BANACH SPACES 65
    Characterization of Normed Spaces (Theorem 4.1.)
    Subspaces and Quotient Spaces
    Isomorphism Theorem
    Open Mapping Theorem
    Closed Graph Theorem
    The Space of Continuous Linear Transformations from X to Y
    Uniform Boundedness Theorem
    Riesz Interpolation Theorem

V. DUAL SPACES 81
The Weak and Strong Topologies
The Weak* Topology
Duality Theorem
The Space S^* of Tempered Distributions
The Conjugate of a Normed Linear Space
Reflexive Spaces
Alaoglu's Theorem

VI. APPLICATIONS TO ANALYSIS 99
Integral Operators
Convolution
Young's Inequality
Reproducing Kernels and Approximate Identities
Poisson Kernels
Gauss Kernel
Green's Functions
Fourier Transform
Inversion Formula
Plancherel Theorem
Tempered Distributions
Hilbert Transform

VII. AXIOMS FOR A MATHEMATICAL
MODEL OF EXPERIMENTAL SCIENCE 125
States and Observables
Initial Axioms
Questions
Question-Valued Measures
Automorphisms and Homomorphisms of the Set of Questions
Time Evolution of the System
Symmetries of the System
Additional Axioms

VIII. HILBERT SPACES 141
     Inner Product Spaces
     Cauchy-Schwarz Inequality
     Orthonormal Sets and Bases
     Bessel's Inequality and Parseval's Equality
     Projection Theorem
     Hilbert Space Direct Sums
     Riesz Representation Theorem for Hilbert Spaces
     Bounded Operators on Hilbert Space
     The Adjoint of a Bounded Operator
     Classes of Bounded Operators
     The Set $\mathcal{P}$ of All Orthogonal Projections

IX. PROJECTION-VALUED MEASURES 165
     Borel Spaces
     Canonical Projection-Valued Measures
     Cyclic Vectors and Separating Vectors
     The Integral with Respect to p of a Simple Function
     The Integral with Respect to p of a Bounded Function
     A "Riesz" Representation Theorem
     The Integral with Respect to p of an Unbounded Function

X. THE SPECTRAL THEOREM OF GELFAND 187
     C^*-Algebras
     Mazur's Theorem
     The Structure Space and the Gelfand Transform
     The Spectrum of an Element
     Spectral Radius and the Norm of the Gelfand Transform
     General Spectral Theorem
     Spectral Theorem for a Normal Operator
     Stone's Theorem

XI. APPLICATIONS OF SPECTRAL THEORY 211
Square Root of a Positive Operator
Polar Decomposition Theorem
Special Subsets of the Spectrum
Compact Operators
Hilbert-Schmidt Operators
Trace Class Operators
Hilbert-Schmidt Norm
Trace Norm
Unbounded Selfadjoint Operators
Spectral Theorem for an Unbounded Selfadjoint Operator
Invariance of the Essential Spectrum under Compact Perturbations
A Characterization of Unbounded Selfadjoint Operators
Stone's Theorem

XII. NONLINEAR FUNCTIONAL ANALYSIS,
INFINITE-DIMENSIONAL CALCULUS 241
The Differential of a Function on a Banach Space
Chain Rule
Mean Value Theorem
First Derivative Test
Continuous Differentiability
Theorem on Mixed Partials (Symmetry of the Second Differential)
Second Derivative Test
Contraction Mapping Theorem
Implicit Function Theorem
Inverse Function Theorem
Foliated Implicit Function Theorem
Differentiable Manifolds
Tangent Vectors
Method of Lagrange Multipliers

BIBLIOGRAPHY 259

INDEX 261

CHAPTER 0

PRELIMINARIES

We include in this preliminary chapter some of the very basic concepts and results of set theory, linear algebra, and topology. We do this so that precise definitions and theorems will be at hand for reference. The exercises given here contain some of the main results. Although they should be routine for the student of this subject, we recommend that they be done carefully. The main theorems of Functional Analysis frequently rely on the Axiom of Choice, and in some cases are equivalent to this axiom from abstract set theory. The version of the Axiom of Choice that is ordinarily used in Functional Analysis is the Hausdorff maximality principle, which we state here without proof.

HAUSDORFF MAXIMALITY PRINCIPLE. *Let S be a nonempty set, and let $<$ denote a partial ordering on S, i.e., a transitive relation on S. Then there exists a maximal linearly ordered subset of S.*

Frequently encountered in our subject is the notion of an infinite product.

DEFINITION. Let I be a set, and for each $i \in I$ let X_i be a set. By the *Cartesian product* of the sets $\{X_i\}$, we mean the set of all functions f defined on I for which $f(i) \in X_i$ for each $i \in I$. We denote this set of functions by $\prod_{i \in I} X_i$ or simply by $\prod X_i$.

Ordinarily, a function $f \in \prod_{i \in I} X_i$ is denoted by $\{x_i\}$, where $x_i = f(i)$.

Fundamental to Functional Analysis are the notions of vector spaces and linear transformations.

DEFINITION. Let F denote either the field $\mathbb{R}$ of real numbers or the field $\mathbb{C}$ of complex numbers. A *vector space* over F is an additive abelian group X, on which the elements of F act by scalar multiplication:

   (1) $a(x + y) = ax + ay$ and $(a + b)x = ax + bx$ for all $x, y \in X$ and $a, b \in F$.
   (2) $1x = x$ for all $x \in X$.

The elements of F are called *scalars*. If $F = \mathbb{R}$, then X is called a *real vector space*, and if $F = \mathbb{C}$, then X is called a *complex vector space*. Obviously, a complex vector space can also be regarded as a real vector space, but not every real vector space is a complex vector space. See Exercise 0.1 below.

A subset Y of a vector space X is called a *subspace* if it is closed under addition and scalar multiplication.

A nonempty finite set $\{x_1, \ldots, x_n\}$ of nonzero elements of a vector space X is called *linearly dependent* if there exist elements $\{a_1, \ldots, a_n\}$ of F, not all 0, such that $\sum_{i=1}^n a_i x_i = 0$. An arbitrary set S of nonzero elements of X is called *linearly dependent* if some nonempty finite subset of S is linearly dependent. A subset $S \subseteq X$ of nonzero vectors is called *linearly independent* if it is not linearly dependent.

A subset B of a vector space X is said to be a *spanning set* for X if every element of X is a finite linear combination of elements of B. A *basis* of X is a linearly independent spanning subset of X.

EXERCISE 0.1. (a) Prove that every vector space has a basis. HINT: The Hausdorff maximality principle.

   (b) If B is a basis of a vector space X, show that each element $x \in X$ can be written **uniquely** as a finite linear combination $x = \sum_{i=1}^n a_i x_i$, where each $x_i \in B$.

   (c) Show that any two bases of a vector space have the same cardinality, i.e., they can be put into 1-1 correspondence.

   (d) Show that the set F^n of all n-tuples $(x_1, x_2, \ldots, x_n)$ of elements of F is a vector space with respect to coordinatewise addition and scalar multiplication.

   (e) Prove that every complex vector space is automatically a real vector space. On the other hand, show that $\mathbb{R}^3$ is a real vector space but that scalar multiplication cannot be extended to $\mathbb{C}$ so that $\mathbb{R}^3$ is a complex vector space. HINT: What could ix possibly be?

DEFINITION. The *dimension* of a vector space X is the cardinality of a basis of X.

DEFINITION. Let I be a set, and let $\{X_i\}$, for $i \in I$, be a collection of vector spaces over F. By the *vector space direct product* $\prod_{i \in I} X_i$, we mean the cartesian product of the sets $\{X_i\}$, together with the operations:

(1) $\{x_i\} + \{y_i\} = \{x_i + y_i\}$
(2) $a\{x_i\} = \{ax_i\}$.

The (algebraic) *direct sum*

$$\bigoplus_{i \in I} X_i$$

is defined to be the subset of $\prod_{i \in I} X_i$ consisting of the elements $\{x_i\}$ for which $x_i = 0$ for all but a finite number of i's.

EXERCISE 0.2. (a) Prove that $\prod_{i \in I} X_i$ is a vector space.
(b) Show that $\bigoplus_{i \in I} X_i$ is a subspace of $\prod_{i \in I} X_i$.
(c) Prove that

$$F^n = \prod_{i \in \{1,\dots,n\}} F = \sum_{i \in \{1,\dots,n\}} F.$$

DEFINITION. A *linear transformation* from a vector space X into a vector space Y is a function $T : X \to Y$ for which

$$T(a_1 x_1 + a_2 x_2) = a_1 T(x_1) + a_2 T(x_2)$$

for all $x_1, x_2 \in X$ and $a_1, a_2 \in F$.

A linear transformation $T : X \to Y$ is called a *linear isomorphism* if it is 1-1 and onto.

By the *kernel* $\ker(T)$ of a linear transformation T, we mean the set of all $x \in X$ for which $T(x) = 0$, and by the *range* of T we mean the set of all elements of Y of the form $T(x)$.

EXERCISE 0.3. Let X and Y be vector spaces, and let B be a basis for X.
(a) Suppose T and S are linear transformations of X into Y. Show that $T = S$ if and only if $T(x) = S(x)$ for every $x \in B$.
(b) For each $b \in B$ let y_b be an element of Y. Show that there exists a (unique) linear transformation $T : X \to Y$ satisfying $T(b) = y_b$ for all $b \in B$.
(c) Let T be a linear transformation of X into Y. Prove that the kernel of T is a subspace of X and the range of T is a subspace of Y.

(d) Let $T : X \to Y$ be a linear isomorphism. Prove that $T^{-1} : Y \to X$ is a linear isomorphism.

DEFINITION. A *linear functional* on a vector space X over F is a linear transformation of X into $F \equiv F^1$.

EXERCISE 0.4. Let f be a linear functional on a vector space X, and let M be the kernel of f.

(a) If x is an element of X, which is not in M, show that every element $y \in X$ can be written uniquely as $y = m + ax$, where $m \in M$ and $a \in F$.

(b) Let f and g be linear functionals on X. Show that f is a nonzero multiple $f = ag$ of g if and only if $\ker(f) = \ker(g)$.

(c) Let T be a linear transformation of a vector space X onto the vector space F^n. Show that there exist elements $\{x_1, \dots, x_n\}$ of X, none of which belongs to $\ker(T)$, such that each element $y \in X$ can be written uniquely as $y = m + \sum_{i=1}^n a_i x_i$, where $m \in \ker(T)$ and each $a_i \in F$.

DEFINITION. If X is a vector space and M is a subspace of X, we define the *quotient space* X/M to be the set of all cosets $x + M$ of M together with the following operations:

$$(x + M) + (y + M) = (x + y) + M,$$

and

$$a(x + M) = ax + M$$

for all $x, y \in X$ and $a \in F$.

EXERCISE 0.5. Let M be a subspace of a vector space X.

(a) Prove that the quotient space X/M is a vector space.

(b) Define $\pi : X \to X/M$ by $\pi(x) = x + M$. Show that π is a linear transformation from X onto X/M. This transformation π is called the *natural map* or *quotient map* of X onto X/M.

(c) If T is a linear transformation of X into a vector space Y, and if $M \subseteq \ker(T)$, show that there exists a unique linear transformation $S : X/M \to Y$ such that $T = S \circ \pi$, where π is the natural map of X onto X/M.

Perhaps the most beautiful aspect of Functional Analysis is in its combining of linear algebra and topology. We give next the fundamental topological ideas that we will need.

DEFINITION. A *topology* on a set X is a collection T of subsets of X satisfying:

(1) $X \in T$.

(2) $\emptyset \in T$.

(3) The intersection of any finite number of elements of T is an element of T.

(4) The union of an arbitrary collection of elements of T is an element of T.

The set X, or the pair (X, T), is called a *topological space.*

The elements of a topology T are called *open* subsets of X, and their complements are called *closed* sets. An open set containing a point $x \in X$ is called an *open neighborhood* of x, and any set that contains an open neighborhood of x is itself called a *neighborhood* of x.

If A is a subset of a topological space (X, T) and x is a point of A, then x is called an *interior point* of A if A contains a neighborhood of x. The *interior* of A is the set of all interior points of A.

If Y is a subset of a topological space (X, T), then the *relative topology on* Y is the collection T' of subsets of Y obtained by intersecting the elements of T with Y. The collection T' is a topology on Y, and the pair (Y, T') is called a *topological subspace* of X.

A subset B of a topology T is called a *base* for T if each element $U \in T$ is a union of elements of B.

A topological space (X, T) is called *second countable* if there exists a countable base B for T.

A topological space (X, T) is called a *Hausdorff* space if for each pair of distinct points $x, y \in X$ there exist open sets $U, V \in T$ such that $x \in U$, $y \in V$, and $U \cap V = \emptyset$. X is called a *regular* topological space if, for each closed set $A \subseteq X$ and each point $x \notin A$, there exist open sets U and V such that $A \subseteq U$, $x \in V$, and $U \cap V = \emptyset$. X is called a *normal* topological space if, for each pair A, B of disjoint closed subsets of X, there exist open sets U, V such that $A \subseteq U$, $B \subseteq V$, and $U \cap V = \emptyset$.

By an *open cover* of a subset Y of a topological space X, we mean a collection U of open subsets of X for which $Y \subseteq \cup_{U \in U} U$. A subset Y of a topological space X is called *compact* if every open cover U of Y has a finite subcover; i.e., there exist finitely many elements $U_1, \ldots, U_n$ of U such that $Y \subseteq \cup_1^n U_i$.

A topological space X is called *σ-compact* if it is a countable union of compact subsets.

A topological space X is called *locally compact* if, for every $x \in X$ and every open set U containing x, U contains a compact neighborhood of x.

A function F from one topological space X into another topological space Y is called *continuous* if $f^{-1}(U)$ is an open subset of X whenever U is an open subset of Y.

A *metric* on a set X is a function $d : X \times X \to \mathbb{R}$ that satisfies:

(1) $d(x, y) \geq 0$ for all $x, y \in X$.
(2) $d(x, y) = 0$ if and only if $x = y$.
(3) (Triangle inequality) $d(x, z) \leq d(x, y) + d(y, z)$ for all $x, y, z \in X$.

If X is a set on which a metric d is defined, then X (or the pair (X, d)) is called a *metric space*.

If d is a metric on a set X, x is an element of X, and $\epsilon > 0$, then the *ball* $B_\epsilon(x)$ of radius ϵ around x is defined to be the set of all $y \in X$ for which $d(x, y) < \epsilon$. A point x is called an *interior* point of a subset A of a metric space (X, d) if there exists an $\epsilon > 0$ such that $B_\epsilon(x) \subseteq A$, and a set A is called *open* relative to a metric d if every point of A is an interior point of A.

The topological space $(X, \mathcal{T})$ is called *metrizable* if there exists a metric d on X for which the elements of $\mathcal{T}$ coincide with the sets that are open sets relative to the metric d.

EXERCISE 0.6. (a) Let $\mathcal{A}$ be a collection of subsets of a set X. Prove that there is a smallest topology $\mathcal{T}$ on X that contains $\mathcal{A}$, and verify that a base for this topology consists of the collection of all sets B of the form

$$B = \cap_{i=1}^n A_i,$$

where each $A_i \in \mathcal{A}$.

(b) Let A be a subset of a topological space $(X, \mathcal{T})$. Prove that the interior of A is an open set. Prove that the intersection of all closed sets containing A is closed. This closed set is called the *closure* of A and is denoted by $\bar{A}$.

(c) Let Y be a subset of a topological space $(X, \mathcal{T})$, and write $\mathcal{T}'$ for the collection of subsets V of Y of the form $V = U \cap Y$ for $U \in \mathcal{T}$. Prove that $\mathcal{T}'$ is a topology on Y.

(d) Let d be a metric on a set X. Show that the collection of all sets that are open relative to d forms a topology on X.

(e) Let X and Y be topological spaces. Prove that a function $f : X \to Y$ is continuous if and only if for every open set $U \subseteq Y$ and every $x \in f^{-1}(U)$ there exists an open set $V \subseteq X$ such that $x \in V$ and $f(V) \subseteq U$.

EXERCISE 0.7. Let X be a set, and let $\{X_i\}$, for i in a set I, be a collection of topological spaces. For each i, let f_i be a map of X into X_i.

(a) Prove that there exists a smallest topology $\mathcal{T}$ on X for which each function f_i is continuous.

(b) Let $\mathcal{T}$ be as in part a. Show that, for each index i and each open subset $U_i \subseteq X_i$, the set $f_i^{-1}(U_i)$ belongs to $\mathcal{T}$.

(c) Let $\mathcal{T}$ be as in part a. Show that, for each finite set $i_1, \ldots, i_n$ of elements of I, and for each n-tuple $U_{i_1}, \ldots, U_{i_n}$, for U_{i_j} an open subset of X_{i_j}, the set

$$\cap_{j=1}^{n} f_{i_j}^{-1}(U_{i_j})$$

is in $\mathcal{T}$.

(d) Let $\mathcal{T}$ be as in part a. Show that each element of $\mathcal{T}$ is a union of sets of the form described in part c; i.e., the sets described in part c form a base for $\mathcal{T}$.

DEFINITION. Let X be a set, and for each i in a set I let f_i be a function from X into a topological space X_i. The smallest topology on X, for which each f_i is continuous, is called the *weak topology generated by the f_i's*.

If $\{X_i\}$, for $i \in I$, is a collection of topological spaces, write

$$X = \prod_{i \in I} X_i,$$

and define $f_i : X \to X_i$ by

$$f_i(\{x_j\}) = x_i.$$

The *product topology* on $X = \prod_{i \in I} X_i$ is defined to be the weak topology generated by the f_i's.

EXERCISE 0.8. Let X be a set, let $\{X_i\}$ for $i \in I$, be a collection of topological spaces, and for each $i \in I$ let f_i be a map of X into X_i. Let $\mathcal{T}$ denote the weak topology on X generated by the f_i's.

(a) Prove that $\mathcal{T}$ is Hausdorff if each X_i is Hausdorff and the functions $\{f_i\}$ separate the points of X. (The f_i's *separate* the points of X if $x \neq y \in X$ implies that there exists an $i \in I$ such that $f_i(x) \neq f_i(y)$.)

(b) Show that $\mathcal{T}$ is second countable if the index set I is countable and each topological space X_i is second countable.

(c) Conclude that the product space $Y = \prod_{i \in I} X_i$ is second countable if I is countable and each X_i is second countable.

(d) Suppose the index set I is countable and that each X_i is metrizable. Prove that $(X, \mathcal{T})$ is metrizable. HINT: Identify I with the set $\{1, 2, \ldots\}$. If d_i denotes the metric on X_i, define d on X by

$$d(x, y) = \sum_{i=1}^{\infty} 2^{-i} \min(1, d_i(f_i(x), f_i(y))),$$

and show that d is a metric whose open sets coincide with the elements of $\mathcal{T}$.

(e) Let Y be the topological product space $Y = \prod_{i \in I} X_i$, and define $F : X \to Y$ by $[F(x)]_i = f_i(x)$. Prove that F is a homeomorphism of $(X, \mathcal{T})$ into Y.

EXERCISE 0.9. (a) Prove that a topological space X is compact if and only if it satisfies the finite intersection property; i.e., if $\mathcal{F}$ is a collection of closed subsets of X, for which the intersection of any finite number of elements of $\mathcal{F}$ is nonempty, then the intersection of all the elements of $\mathcal{F}$ is nonempty.

(b) Prove that a compact Hausdorff space is normal.

(c) Prove that a regular space, having a countable base, is normal.

(d) Prove Urysohn's Lemma: If X is a normal topological space, and if A and B are nonempty disjoint closed subsets of X, then there exists a continuous function $f : X \to [0, 1]$ such that $f(A) = \{0\}$ and $f(B) = \{1\}$.

(e) Let X be a regular space having a countable base. Show that there exists a sequence $\{f_n\}$ of continuous real-valued functions on X, such that for each closed set $A \subseteq X$ and each point $x \notin A$, there exists an n for which $f_n(x) \notin f_n(A)$. HINT: For each pair U, V of elements of the countable base, for which $U \subseteq \bar{U} \subset V$, use Urysohn's lemma on the sets $\bar{U}$ and $\tilde{V}$, where $\tilde{V}$ denotes the complement of V. Conclude that the topology on X coincides with the weak topology generated by the resulting f_n's.

(f) Prove that a regular space X, having a countable base, is metrizable. HINT: Use part e to construct a homeomorphism between X and a subset of a countable product of real lines.

(g) Prove that a locally compact Hausdorff space is regular and hence that a locally compact, second countable, Hausdorff space is metrizable.

DEFINITION. Let $(X, \mathcal{T})$ be a topological space, and let f be a function from X onto a set Y. The largest topology $\mathcal{Q}$ on Y for which f is continuous is called the quotient topology on Y.

EXERCISE 0.10. Let $(X, \mathcal{T})$ be a topological space, let $f : X \to Y$ be a map of X onto a set Y, and let $\mathcal{Q}$ be the quotient topology on Y.

(a) Prove that a subset $U \subseteq Y$ belongs to $\mathcal{Q}$ if and only if $f^{-1}(U)$ belongs to $\mathcal{T}$. That is, $\mathcal{Q} = \{U \subseteq Y : f^{-1}(U) \in \mathcal{T}\}$.

(b) Suppose Z is a topological space and that g is a function from $(Y, \mathcal{Q})$ into Z. Prove that g is continuous if and only if $g \circ f$ is continuous from $(X, \mathcal{T})$ into Z.

CHAPTER I

THE RIESZ REPRESENTATION THEOREM

We begin our study by identifying certain special kinds of linear functionals on certain special vector spaces of functions. We describe these linear functionals in terms of more familiar mathematical objects, i.e., as integrals against measures. We have labeled Theorem 1.3 as the Riesz Representation Theorem. However, each of Theorems 1.2, 1.3, 1.4 and 1.5 is often referred to by this name, and a knowledge of this nontrivial theorem, or set of theorems, is fundamental to our subject. Theorem 1.1 is very technical, but it is the cornerstone of this chapter.

DEFINITION. A *vector lattice* of functions on a set X is a vector space L of real-valued functions on X which is closed under the binary operations of maximum and minimum. That is:

(1) $f, g \in L$ and $\alpha, \beta \in \mathbb{R}$ implies that $\alpha f + \beta g \in L$.
(2) $f, g \in L$ implies that $\max(f, g) \in L$ and $\min(f, g) \in L$.

REMARKS. The set of all continuous real-valued functions on a topological space X clearly forms a vector lattice, indeed the prototypical one. A nontrivial vector lattice certainly contains some nonnegative functions (taking maximum of f and 0). If a vector lattice does not contain any nonzero constant function, it does not follow that the minimum of an $f \in L$ and the constant function 1 must belong to L. The set of all scalar multiples of a fixed positive nonconstant function is a counterexample.

Stone's axiom for a vector lattice L is as follows: If f is a nonnegative function in L, then $\min(f, 1)$ is an element of L.

11

EXERCISE 1.1. Let L be a vector lattice of functions on a set X, and suppose L satisfies Stone's axiom.

(a) Show that $\min(f, c) \in L$ whenever f is a nonnegative function in L and $c \geq 0$.

(b) (A Urysohn-type property) Let E and F be disjoint subsets of X, $0 \leq a < b$, and let $f \in L$ be a nonnegative function such that $f(x) \geq b$ on F and $f(x) \leq a$ on E. Show that there exists an element $g \in L$ such that $0 \leq g(x) \leq 1$ for all $x \in X$, $g(x) = 0$ on E, and $g(x) = 1$ on F.

(c) Let $0 \leq a < b < c < d$ be real numbers, let $f \in L$ be nonnegative, and define $E = f^{-1}([0, a])$, $F = f^{-1}([b, c])$, and $G = f^{-1}([d, \infty))$. Show that there exists an element g of L such that $0 \leq bg(x) \leq f(x)$ for all $x \in X$, $g(x) = 1$ on F, and $g(x) = 0$ on $E \cup G$.

(d) Let μ be a measure defined on a σ-algebra of subsets of the set X, and suppose $L = L^1(\mu)$ is the set of all (absolutely) integrable real-valued functions on X with respect to μ. Show that L is a vector lattice that satisfies Stone's axiom.

(e) Let μ and L be as in part d. Define $\phi : L \to \mathbb{R}$ by $\phi(f) = \int f \, d\mu$. Prove that ϕ is a positive linear functional on L, i.e., ϕ is a linear functional for which $\phi(f) \geq 0$ whenever $f(x) \geq 0$ for all $x \in X$.

We come now to our fundamental representation theorem for linear functionals.

THEOREM 1.1. *Let L be a vector lattice on a set X, and assume that L satisfies Stone's axiom, i.e., that if f is a nonnegative function in L, then $\min(f, 1) \in L$. Suppose I is a linear functional on the vector space L that satisfies:*

(1) *$I(f) \geq 0$ whenever $f(x) \geq 0$ for all $x \in X$. (I is a positive linear functional.)*

(2) *Suppose $\{f_n\}$ is a sequence of nonnegative elements of L, which increases pointwise to an element f of L, i.e., $f(x) = \lim f_n(x)$ for every x, and $f_n(x) \leq f_{n+1}(x)$ for every x and n. Then $I(f) = \lim I(f_n)$. (I satisfies the monotone convergence property.)*

Then there exists a (not necessarily finite) measure μ defined on a σ-algebra $\mathcal{M}$ of subsets of X such that every $f \in L$ is μ-measurable, μ-integrable, and

$$I(f) = \int f \, d\mu.$$

PROOF. We begin by defining an outer measure μ^* on all subsets of

X. Thus, if $E \subseteq X$, put

$$\mu^*(E) = \inf \sum I(h_m),$$

where the infimum is taken over all sequences $\{h_m\}$ of nonnegative functions in L for which $\sum h_m(x) \geq 1$ for each $x \in E$. Note that if, for some set E, no such sequence $\{h_m\}$ exists, then $\mu^*(E) = \infty$, the infimum over an empty set being $+\infty$. In particular, if L does not contain the constant function 1, then $\mu^*(X)$ could be ∞, although not necessarily. See Exercise 1.2 below.

It follows routinely that μ^* is an outer measure. Again see Exercise 1.2 below.

We let μ be the measure generated by μ^*, i.e., μ is the restriction of μ^* to the σ-algebra $\mathcal{M}$ of all μ^*-measurable subsets of X. We wish to show that each $f \in L$ is μ-measurable, μ-integrable, and then that $I(f) = \int f \, d\mu$. Since L is a vector lattice, and both I and $\int \cdot \, d\mu$ are positive linear functionals on L, we need only verify the above three facts for nonnegative functions $f \in L$.

To prove that a nonnegative $f \in L$ is μ-measurable, it will suffice to show that each set $f^{-1}[a, \infty)$, for $a > 0$, is μ^* measurable; i.e., we must show that for any $A \subseteq X$,

$$\mu^*(A) \geq \mu^*(A \cap f^{-1}[a, \infty)) + \mu^*(A \cap \widetilde{f^{-1}[a, \infty)}).$$

We first make the following observation.

Suppose $A \subseteq X$, $0 < a < b$, E is a subset of X for which $f(x) \leq a$ if $x \in E$, and F is a subset of X for which $f(x) \geq b$ if $x \in F$. Then

$$\mu^*(A \cap (E \cup F)) \geq \mu^*(A \cap E) + \mu^*(A \cap F).$$

Indeed, let g be the element of L defined by

$$g = \frac{\min(f, b) - \min(f, a)}{b - a}.$$

Then $g = 0$ on E, and $g = 1$ on F. If $\epsilon > 0$ is given, and $\{h_m\}$ is a sequence of nonnegative elements of L for which $\sum h_m(x) \geq 1$ on $A \cap (E \cup F)$, and $\sum I(h_m) < \mu^*(A \cap (E \cup F)) + \epsilon$, set $f_m = \min(h_m, g)$ and $g_m = h_m - \min(h_m, g)$. Then:

$$h_m = f_m + g_m$$

on X,
$$\sum f_m(x) \geq 1$$
for $x \in A \cap F$, and
$$\sum g_m(x) \geq 1$$
for $x \in A \cap E$. Therefore:

$$\mu^*(A \cap (E \cup F)) + \epsilon \geq \sum I(h_m)$$
$$= \sum I(f_m) + \sum I(g_m)$$
$$\geq \mu^*(A \cap F) + \mu^*(A \cap E).$$

It follows now by induction that if $\{I_1, ..., I_n\}$ is a finite collection of disjoint half-open intervals $(a_j, b_j]$, with $0 < b_1$ and $b_j < a_{j+1}$ for $1 \leq j < n$, and if $E_j = f^{-1}(I_j)$, then

$$\mu^*(A \cap (\cup E_j)) \geq \sum \mu^*(A \cap E_j)$$

for any subset A of X. In fact, using the monotonicity of the outer measure μ^*, the same assertion is true for any countable collection $\{I_j\}$ of such disjoint half-open intervals. See Exercise 1.3.

Now, Let A be an arbitrary subset of X, and let $a > 0$ be given. Write $E = f^{-1}[a, \infty)$. We must show that

$$\mu^*(A \cap E) + \mu^*(A \cap \tilde{E}) \leq \mu^*(A).$$

We may assume that $\mu^*(A)$ is finite, for otherwise the desired inequality is obvious. Let $\{c_1, c_2, ...\}$ be a strictly increasing sequence of positive numbers that converges to a. We write the interval $(-\infty, a)$ as the countable union $\cup_{j=0}^{\infty} I_j$ of the disjoint half-open intervals $\{I_j\}$, where $I_0 = (-\infty, c_1]$, and for $j > 0$, $I_j = (c_j, c_{j+1}]$, whence

$$\tilde{E} = \cup_{j=0}^{\infty} E_j,$$

where $E_j = f^{-1}(I_j)$. Also, if we set $F_k = \cup_{j=0}^{k} E_j$, then $\tilde{E}$ is the increasing union of the F_k's. Then, using Exercise 1.3, we have:

$$\mu^*(A \cap (\cup_{j=0}^{\infty} E_{2j})) \geq \sum_{j=0}^{\infty} \mu^*(A \cap E_{2j}),$$

whence the infinite series on the right is summable.

Similarly, the infinite series $\sum_{j=0}^{\infty} \mu^*(A \cap E_{2j+1})$ is summable. Therefore,

$$
\begin{aligned}
\mu^*(A \cap F_k) &\leq \mu^*(A \cap \tilde{E}) \\
&= \mu^*((A \cap F_k) \cup (A \cap (\cup_{j=k+1}^{\infty} E_j))) \\
&\leq \mu^*(A \cap F_k) + \sum_{j=k+1}^{\infty} \mu^*(A \cap E_j),
\end{aligned}
$$

and this shows that $\mu^*(A \cap \tilde{E}) = \lim_k \mu^*(A \cap F_k)$.

So, recalling that $F_k = f^{-1}(-\infty, c_{k+1}]$, we have

$$
\begin{aligned}
\mu^*(A \cap E) + \mu^*(A \cap \tilde{E}) &= \mu^*(A \cap E) + \lim_k \mu^*(A \cap F_k) \\
&= \lim_k (\mu^*(A \cap E) + \mu^*(A \cap F_k)) \\
&\leq \lim_k \mu^*(A \cap (E \cup F_k)) \\
&\leq \mu^*(A),
\end{aligned}
$$

as desired. Therefore, f is μ-measurable for every $f \in L$.

It remains to prove that each $f \in L$ is μ-integrable, and that $I(f) = \int f \, d\mu$. It will suffice to show this for f's which are nonnegative, bounded, and 0 outside a set of finite μ measure. See Exercise 1.4. For such an f, let ϕ be a nonnegative measurable simple function, with $\phi(x) \geq f(x)$ for all x, and such that ϕ is 0 outside a set of finite measure. (We will use the fact from measure theory that there exists a sequence $\{\phi_n\}$ of such simple functions for which $\int f \, d\mu = \lim \int \phi_n \, d\mu$.) Write $\phi = \sum_{i=1}^{k} a_i \chi_{E_i}$, where each $a_i \geq 0$, and let $\epsilon > 0$ be given. For each i, let $\{h_{i,m}\}$ be a sequence of nonnegative elements in L for which $\sum_m h_{i,m}(x) \geq 1$ on E_i, and

$\sum_m I(h_{i,m}) < \mu^*(E_i) + \epsilon$. Then

$$\int \phi \, d\mu + \epsilon \sum_{i=1}^{k} a_i = \sum_{i=1}^{k} a_i \mu(E_i) + \epsilon \sum_{i=1}^{k} a_i$$

$$> \sum_{i=1}^{k} a_i \sum_{m} I(h_{i,m})$$

$$= \sum_{m} \sum_{i=1}^{k} a_i I(h_{i,m})$$

$$= \lim_{M} \sum_{m=1}^{M} \sum_{i=1}^{k} a_i I(h_{i,m})$$

$$= \lim_{M} I(h_M),$$

where

$$h_M = \sum_{m=1}^{M} \sum_{i=1}^{k} a_i h_{i,m}.$$

Observe that $\{h_M\}$ is an increasing sequence of nonnegative elements of L, and that $\lim h_M(x) \geq \phi(x)$ for all $x \in X$, whence the sequence $\{\min(h_M, f)\}$ increases pointwise to f. Therefore, by the monotone convergence property of I, we have that

$$\int \phi \, d\mu + \epsilon \sum_{i=1}^{k} a_i \geq \lim_{M} I(h_M)$$

$$\geq \lim_{M} I(\min(h_M, f))$$

$$= I(f),$$

showing that $\int \phi \, d\mu \geq I(f)$, for all such simple functions ϕ. It follows then that $\int f \, d\mu \geq I(f)$.

To show the reverse inequality, we may suppose that $0 \leq f(x) < 1$ for all x, since both I and $\int \cdot \, d\mu$ are linear. For each positive integer n and each $0 \leq i < 2^n$, define the set $E_{i,n}$ by

$$E_{i,n} = f^{-1}([i/2^n, (i+1)/2^n)),$$

and then a simple function ϕ_n by

$$\phi_n = \sum_{i=0}^{2^n-1} (i/2^n) \chi_{E_{i,n}}.$$

Using Exercise 1.1 part c, we choose, for each $0 \leq i < 2^n$ and each $m > 2^{n+1}$, a function $g_{i,m}$ satisfying:

(1) For $x \in f^{-1}([i/2^n, ((i+1)/2^n) - 1/m))$,

$$g_{i,m}(x) = i/2^n.$$

(2) For $x \in f^{-1}([0, (i/2^n) - 1/2m))$ and $x \in f^{-1}([((i+1)/2^n) - 1/2m, 1])$,

$$g_{i,m}(x) = 0.$$

(3) For all x,

$$g_{i,m}(x) \leq f(x).$$

Then

$$\mu(E_{i,n}) = \lim_m \mu(f^{-1}([i/2^n, (i+1)/2^n - 1/m))).$$

And,

$$\sum_{i=0}^{2^n-1} (i/2^n)\mu(f^{-1}([i/2^n, (i+1)/2^n - 1/m))) \leq \sum_{i=0}^{2^n-1} I(g_{i,m})$$

$$= I(h_m),$$

where

$$h_m = \sum_{i=0}^{2^n-1} g_{i,m}.$$

Observe that $h_M(x) \leq f(x)$ for all x. It follows that

$$\int \phi_n \, d\mu = \sum_{i=0}^{2^n-1} (i/2^n)\mu(E_{i,n})$$

$$= \lim_m \sum_{i=0}^{2^n-1} (i/2^n)\mu(f^{-1}([i/2^n, (i+1)/2^n - 1/m)))$$

$$\leq \limsup_m I(h_m)$$

$$\leq I(f),$$

whence, by letting n tend to ∞, we see that $\int f \, d\mu \leq I(f)$. The proof of the theorem is now complete.

EXERCISE 1.2. (a) Give an example of a vector lattice L of functions on a set X, such that the constant function 1 does not belong to L, but for which there exists a sequence $\{h_n\}$ of nonnegative elements of L satisfying $\sum h_n(x) \geq 1$ for all $x \in X$.

(b) Verify that the μ^* in the preceding proof is an outer measure on X by showing that:

(1) $\mu^*(\emptyset) = 0$.

(2) If E and F are subsets of X, with E contained in F, then $\mu^*(E) \leq \mu^*(F)$.

(3) μ^* is *countably subadditive*, i.e.,

$$\mu^*(\cup E_n) \leq \sum \mu^*(E_n)$$

for every sequence $\{E_n\}$ of subsets of X.

HINT: To prove the countable subadditivity, assume that each $\mu^*(E_n)$ is finite. Then, given any $\epsilon > 0$, let $\{h_{n,i}\}$ be a sequence of nonnegative functions in L for which $\sum_i h_{n,i}(x) \geq 1$ for all $x \in E_n$ and for which $\sum_i I(h_{n,i}) \leq \mu^*(E_n) + \epsilon/2^n$.

EXERCISE 1.3. Let $\{I_1, I_2, ...\}$ be a countable collection of half-open intervals $(a_j, b_j]$, with $0 < b_1$ and $b_j < a_{j+1}$ for all j. Let f be a nonnegative element of the lattice L of the preceding theorem, and set $E_j = f^{-1}(I_j)$. Show that for each $A \subseteq X$ we have

$$\mu^*(A \cap (\cup E_j)) = \sum \mu^*(A \cap E_j).$$

HINT: First show this, by induction, for a finite sequence $I_1, \ldots, I_n$, and then verify the general case by using the properties of the outer measure.

EXERCISE 1.4. Let L be the lattice of the preceding theorem.

(a) Show that there exist sets of finite μ-measure. In fact, if f is a nonnegative element of L, show that $f^{-1}([\epsilon, \infty))$ has finite measure for every positive ϵ.

(b) Let $f \in L$ be nonnegative. Show that there exists a sequence $\{f_n\}$ of bounded nonnegative elements of L, each of which is 0 outside some set of finite μ-measure, which increases to f. HINT: Use Stone's axiom.

(c) Conclude that, if $I(f) = \int f \, d\mu$ for every $f \in L$ that is bounded, nonnegative, and 0 outside a set of finite μ-measure, then $I(f) = \int f \, d\mu$ for every $f \in L$.

REMARK. One could imagine that all linear functionals defined on a vector lattice of functions on a set X are related somehow to integration over X. The following exercise shows that this is not the case; that is, some extra hypotheses on the functional I are needed.

EXERCISE 1.5. (a) Let X be the set of positive integers, and let L be the space of all functions f (sequences) on X for which $\lim_{n \to \infty} f(n)$ exists. Prove that L is a vector lattice that satisfies Stone's axiom.

(b) Let X and L be as in part a, and define $I : L \to \mathbb{R}$ by $I(f) = \lim_{n \to \infty} f(n)$. Prove that I is a positive linear functional.

(c) Let I be the positive linear functional from part b. Prove that there exists no measure μ on the set X for which $I(f) = \int f \, d\mu$ for all $f \in L$. HINT: If there were such a measure, there would have to exist a sequence $\{\mu_n\}$ such that $I(f) = \sum f(n)\mu_n$ for all $f \in L$. Show that each μ_n must be 0, and that this would lead to a contradiction.

(d) Let X, L, and I be as in part b. Verify by giving an example that I fails to satisfy the monotone convergence property of Theorem 1.1.

DEFINITION. If Δ is a Hausdorff topological space, then the smallest σ-algebra $\mathcal{B}$ of subsets of Δ, which contains all the open subsets of Δ, is called the σ-algebra of *Borel* sets. A measure which is defined on this σ-algebra, is called a *Borel* measure. A function f from Δ into another topological space Δ' is called a *Borel function* if $f^{-1}(U)$ is a Borel subset of Δ whenever U is an open (Borel) subset of Δ'.

A real-valued (or complex-valued) function f on Δ is said to have *compact support* if the closure of the set of all $x \in \Delta$ for which $f(x) \neq 0$ is compact. The set of all continuous functions having compact support on Δ is denoted by $C_c(\Delta)$.

A real-valued (or complex-valued) function f on Δ is said to *vanish at infinity* if, for each $\epsilon > 0$, the set of all $x \in \Delta$ for which $|f(x)| \geq \epsilon$ is compact. The set of all continuous real-valued functions vanishing at infinity on Δ is denoted here by $C_0(\Delta)$. sometimes, $C_0(\Delta)$ denotes the complex vector space of all continuous complex-valued functions on Δ that vanish at ∞. Hence, the context in which this symbol occurs dictates which meaning it has.

If Δ is itself compact, then every continuous function vanishes at infinity, and we write $C(\Delta)$ for the space of all continuous real-valued (complex-valued) functions on Δ. That is, if Δ is compact, then $C(\Delta) = C_0(\Delta)$.

EXERCISE 1.6. (a) Prove that a second countable locally compact Hausdorff space Δ is metrizable. (See Exercise 0.9.) Conclude that if

K is a compact subset of a second countable locally compact Hausdorff space Δ, then there exists an element $f \in C_c(\Delta)$ that is identically 1 on K.

(b) Let Δ be a locally compact Hausdorff space. Show that every element of $C_c(\Delta)$ is a Borel function, and hence is μ-measurable for every Borel measure μ on Δ.

(c) Show that, if Δ is second countable, Hausdorff, and locally compact, then the σ-algebra of Borel sets coincides with the smallest σ-algebra that contains all the compact subsets of Δ.

(d) If Δ is second countable, Hausdorff, and locally compact, show that the σ-algebra $\mathcal{B}$ of Borel sets coincides with the smallest σ-algebra $\mathcal{M}$ of subsets of Δ for which each $f \in C_c(\Delta)$ satisfies $f^{-1}(U) \in \mathcal{M}$ whenever U is open in $\mathbb{R}$.

(e) Suppose μ and ν are finite Borel measures on a second countable, locally compact, Hausdorff space Δ, and assume that $\int f \, d\mu = \int f \, d\nu$ for every $f \in C_c(\Delta)$. Prove that $\mu = \nu$. HINT: Show that μ and ν agree on compact sets, and hence on all Borel sets.

(f) Prove that a second countable locally compact Hausdorff space Δ is σ-compact. In fact, show that Δ is the increasing union $\cup K_n$ of a sequence of compact subsets $\{K_n\}$ of Δ such that K_n is contained in the interior of K_{n+1}. Note also that this implies that every closed subset F of Δ is the increasing union of a sequence of compact sets.

EXERCISE 1.7. Prove Dini's Theorem: If Δ is a compact topological space and $\{f_n\}$ is a sequence of continuous real-valued functions on Δ that increases monotonically to a continuous function f, then $\{f_n\}$ converges uniformly to f on Δ.

THEOREM 1.2. *Let Δ be a second countable locally compact Hausdorff space. Let I be a positive linear functional on $C_c(\Delta)$. Then there exists a unique Borel measure μ on Δ such that, for all $f \in C_c(\Delta)$, f is μ-integrable and $I(f) = \int f \, d\mu$.*

PROOF. Of course $C_c(\Delta)$ is a vector lattice that satisfies Stone's axiom. The given linear functional I is positive, so that this theorem will follow immediately from Theorem 1.1 and Exercise 1.6 if we show that I satisfies the monotone convergence property. Thus, let $\{f_n\}$ be a sequence of nonnegative functions in $C_c(\Delta)$ that increases monotonically to an element $f \in C_c(\Delta)$. If K denotes a compact set such that $f(x) = 0$ for $x \notin K$, Then $f_n(x) = 0$ for all $x \notin K$ and for all n. We let g be a nonnegative element of $C_c(\Delta)$ for which $g(x) = 1$ on K. On the compact set K, the sequence $\{f_n\}$ is converging monotonically to the continuous function f, whence, by Dini's Theorem, this convergence is uniform. Therefore, given an $\epsilon > 0$, there exists an N such that

$$f(x) - f_n(x) = |f(x) - f_n(x)| < \epsilon$$

for all x if $n \geq N$. Hence $f - f_n \leq \epsilon g$ everywhere on X, whence

$$|I(f - f_n)| = I(f - f_n) \leq I(\epsilon g) = \epsilon I(g).$$

Therefore $I(f) = \lim I(f_n)$, as desired.

DEFINITION. If f is a bounded real-valued function on a set X, we define the *supremum norm* of f, denoted by $\|f\|$, or $\|f\|_\infty$, by

$$\|f\| = \|f\|_\infty = \sup_{x \in X} |f(x)|.$$

A linear functional ϕ on a vector space E of bounded functions is called a *bounded linear functional* if there exists a positive constant M such that $|\phi(f)| \leq M\|f\|$ for all $f \in E$.

THEOREM 1.3. (Riesz Representation Theorem) *Suppose Δ is a second countable locally compact Hausdorff space and that I is a positive linear functional on $C_0(\Delta)$. Then there exists a unique finite Borel measure μ on Δ such that*

$$I(f) = \int f \, d\mu$$

for every $f \in C_0(\Delta)$. Further, I is a bounded linear functional on $C_0(\Delta)$. Indeed, $|I(f)| \leq \mu(\Delta)\|f\|_\infty$.

PROOF. First we show that the positive linear functional I on the vector lattice $C_0(\Delta)$ satisfies the monotone convergence property. Thus, let $\{f_n\}$ be a sequence of nonnegative functions in $C_0(\Delta)$, which increases to an element f, and let $\epsilon > 0$ be given. Choose a compact subset $K \subseteq \Delta$ such that $f(x) \leq \epsilon^2$ if $x \notin K$, and let $g \in C_0(\Delta)$ be nonnegative and such that $g = 1$ on K. Again, by Dini's Theorem, there exists an N such that

$$|f(x) - f_n(x)| \leq \epsilon/(1 + I(g))$$

for all $x \in K$ and all $n \geq N$. For $x \notin K$, we have:

$$\begin{aligned} |f(x) - f_n(x)| &= f(x) - f_n(x) \\ &\leq f(x) \\ &= (\sqrt{f(x)})^2 \\ &\leq \epsilon\sqrt{f(x)}, \end{aligned}$$

so that, for all $x \in \Delta$ and all $n \geq N$, we have

$$|f(x) - f_n(x)| \leq (\epsilon/(1 + I(g)))g(x) + \epsilon\sqrt{f(x)}.$$

Therefore,

$$|I(f) - I(f_n)| = I(f - f_n) \leq \epsilon(1 + I(\sqrt{f})).$$

This proves that $I(f) = \lim I(f_n)$, as desired.

Using Theorem 1.1 and Exercise 1.6, let μ be the unique Borel measure on Δ for which $I(f) = \int f \, d\mu$ for every $f \in C_0(\Delta)$. We show next that there exists a positive constant M such that $|I(f)| \leq M\|f\|_\infty$ for each $f \in C_0(\Delta)$; i.e., that I is a bounded linear functional on $C_0(\Delta)$. If there were no such M, there would exist a sequence $\{f_n\}$ of nonnegative elements of $C_0(\Delta)$ such that $\|f_n\|_\infty = 1$ and $I(f_n) \geq 2^n$ for all n. Then, defining $f_0 = \sum_n f_n/2^n$, we have that $f_0 \in C_0(\Delta)$. (Use the Weierstrass M-test.) On the other hand, since I is a positive linear functional, we see that

$$I(f_0) \geq \sum_{n=1}^{N} I(f_n) \geq N$$

for all n, which is a contradiction. Therefore, I is a bounded linear functional, and we let M be a fixed positive constant satisfying $|I(f)| \leq M\|f\|_\infty$ for all $f \in C_0(\Delta)$.

Observe next that if K is a compact subset of Δ, then there exists a nonnegative function $f \in C_0(\Delta)$ that is identically 1 on K and ≤ 1 everywhere on Δ. Therefore,

$$\mu(K) \leq \int f \, d\mu = I(f) \leq M\|f\|_\infty = M.$$

Because Δ is second countable and locally compact, it is σ-compact, i.e., the increasing union $\cup K_n$ of a sequence of compact sets $\{K_n\}$. Hence, $\mu(\Delta) = \lim \mu(K_n) \leq M$, showing that μ is a finite measure. Then, $|I(f)| = |\int f \, d\mu| \leq \mu(\Delta)\|f\|_\infty$, and this completes the proof.

THEOREM 1.4. *Let Δ be a second countable locally compact Hausdorff space, and let ϕ be a bounded linear functional on $C_0(\Delta)$. That is, suppose there exists a positive constant M for which $|\phi(f)| \leq M\|f\|_\infty$ for all $f \in C_0(\Delta)$. Then ϕ is the difference $\phi_1 - \phi_2$ of two positive linear functionals ϕ_1 and ϕ_2, whence there exists a unique finite signed Borel measure μ such that $\phi(f) = \int f \, d\mu$ for all $f \in C_0(\Delta)$.*

PROOF. For f a nonnegative element in $C_0(\Delta)$, define $\phi_1(f)$ by

$$\phi_1(f) = \sup_g \phi(g),$$

where the supremum is taken over all nonnegative functions $g \in C_0(\Delta)$ for which $0 \leq g(x) \leq f(x)$ for all x. Define $\phi_1(f)$, for an arbitrary element $f \in C_0(\Delta)$, by $\phi_1(f) = \phi_1(f_+) - \phi_1(f_-)$, where $f_+ = \max(f, 0)$ and $f_- = -\min(f, 0)$. It follows from Exercise 1.8 below that ϕ_1 is well-defined and is a linear functional on $C_0(\Delta)$. Since the 0 function is one of the g's over which we take the supremum when evaluating $\phi_1(f)$ for f a nonnegative function, we see that ϕ_1 is a positive linear functional. We define ϕ_2 to be the difference $\phi_1 - \phi$. Clearly, since f itself is one of the g's over which we take the supremum when evaluating $\phi_1(f)$ for f a nonnegative function, we see that ϕ_2 also is a positive linear functional, and the remainder of the proof then follows from the Riesz Representation Theorem.

EXERCISE 1.8. Let L be a vector lattice of bounded functions on a set Δ, and let ϕ be a bounded linear functional on L. That is, suppose that M is a positive constant for which $|\phi(f)| \leq M\|\phi\|_\infty$ for all $f \in L$. For each nonnegative $f \in L$ define, in analogy with the preceding proof,

$$\phi_1(f) = \sup_g \phi(g),$$

where the supremum is taken over all $g \in L$ for which $0 \leq g(x) \leq f(x)$ for all $x \in \Delta$.

(a) If f is a nonnegative element of L, show that $\phi_1(f)$ is a finite real number.

(b) If f and f' are two nonnegative functions in L, show that $\phi_1(f + f') = \phi_1(f) + \phi_1(f')$.

(c) For each real-valued $f = f_+ - f_- \in L$, define $\phi_1(f) = \phi_1(f_+) - \phi_1(f_-)$. Suppose g and h are nonnegative elements of L and that $f = g - h$. Prove that $\phi_1(f) = \phi_1(g) - \phi_1(h)$. HINT: $f_+ + h = g + f_-$.

(d) Prove that ϕ_1, as defined in part c, is a well-defined positive linear functional on L.

EXERCISE 1.9. Let Δ be a locally compact, second countable, Hausdorff space, and let $U_1, U_2, \ldots$ be a countable basis for the topology on Δ for which the closure $\overline{U_n}$ of U_n is compact for every n. Let C be the set of all pairs (n, m) for which $\overline{U_n} \subseteq U_m$, and for each $(n, m) \in C$ let $f_{n,m}$ be a continuous function from Δ into $[0, 1]$ that is 1 on $\overline{U_n}$ and 0 on the complement $\widetilde{U_m}$ of U_m.

(a) Show that each $f_{n,m}$ belongs to $C_0(\Delta)$ and that the set of $f_{n,m}$'s separate the points of Δ.

(b) Let A be the smallest algebra of functions containing all the $f_{n,m}$'s. Show that A is uniformly dense in $C_0(\Delta)$. HINT: Use the Stone-Weierstrass Theorem.

(c) Prove that there exists a countable subset D of $C_0(\Delta)$ such that every element of $C_0(\Delta)$ is the uniform limit of a sequence of elements of D. (That is, $C_0(\Delta)$ is a separable metric space with respect to the metric d given by $d(f,g) = \|f - g\|_\infty$.)

EXERCISE 1.10. (a) Define I on $C_c(\mathbb{R})$ by $I(f) = \int f(x)\,dx$. Show that I is a positive linear functional which is not a bounded linear functional.

(b) Show that there is no way to extend the positive linear functional I of part a to all of $C_0(\mathbb{R})$ so that the extension is still a positive linear functional.

EXERCISE 1.11. Let X be a complex vector space, and let f be a complex linear functional on X. Write $f(x) = u(x) + iv(x)$, where $u(x)$ and $v(x)$ are the real and imaginary parts of $f(x)$.

(a) Show that u and v are real linear functionals on the real vector space X.

(b) Show that $u(ix) = -v(x)$, and $v(ix) = u(x)$. Conclude that a complex linear functional is completely determined by its real part.

(c) Suppose a is a real linear functional on the complex vector space X. Define $g(x) = a(x) - ia(ix)$. Prove that g is a complex linear functional on X.

DEFINITION. Let S be a set and let $\mathcal{B}$ be a σ-algebra of subsets of S. By a *finite complex measure* on $\mathcal{B}$ we mean a mapping $\mu : \mathcal{B} \to \mathbb{C}$ that satisfies:

(1) There exists a constant M such that $|\mu(E)| \leq M$ for all $E \in \mathcal{B}$.
(2) $\mu(\emptyset) = 0$.
(3) If $\{E_n\}$ is a sequence of pairwise disjoint elements of $\mathcal{B}$, then the series $\sum \mu(E_n)$ is absolutely summable, and

$$\mu(\cup E_n) = \sum \mu(E_n).$$

EXERCISE 1.12. Let μ be a finite complex measure on a σ-algebra $\mathcal{B}$ of subsets of a set S.

(a) Write the complex-valued function μ on $\mathcal{B}$ as $\mu_1 + i\mu_2$, where μ_1 and μ_2 are real-valued functions. Show that both μ_1 and μ_2 are finite signed measures on $\mathcal{B}$. Show also that $\bar{\mu} = \mu_1 - i\mu_2$ is a finite complex measure on $\mathcal{B}$.

(b) Let X denote the complex vector space of all bounded complex-valued $\mathcal{B}$-measurable functions on S. If $f \in X$, define

$$\int f \, d\mu = \int f \, d\mu_1 + i \int f \, d\mu_2.$$

Prove that the assignment $f \to \int f \, d\mu$ is a linear functional on X and that there exists a constant M such that

$$\left| \int f \, d\mu \right| \leq M \|f\|_\infty$$

for all $f \in X$.

(c) Show that

$$\int f \, d\bar{\mu} = \overline{\int \bar{f} \, d\mu}.$$

HINT: Write $\mu = \mu_1 + i\mu_2$ and $f = u + iv$.

THEOREM 1.5. (Riesz Representation Theorem, Complex Version) Let Δ be a second countable locally compact Hausdorff space, and denote now by $C_0(\Delta)$ the complex vector space of all continuous complex-valued functions on Δ that vanish at infinity. Suppose ϕ is a linear functional on $C_0(\Delta)$ into the field $\mathbb{C}$, and assume that ϕ is a bounded linear functional, i.e., that there exists a positive constant M such that $|\phi(f)| \leq M \|f\|_\infty$ for all $f \in C_0(\Delta)$. Then there exists a unique finite complex Borel measure μ on Δ such that $\phi(f) = \int f \, d\mu$ for all $f \in C_0(\Delta)$. See the preceding exercise.

EXERCISE 1.13. (a) Prove Theorem 1.5. HINT: Write ϕ in terms of its real and imaginary parts ψ and η. Show that each of these is a bounded real-valued linear functional on the real vector space $C_0(\Delta)$ of all real-valued continuous functions on Δ that vanish at infinity, and that

$$\psi(f) = \int f \, d\mu_1$$

and

$$\eta(f) = \int f \, d\mu_2$$

for all real-valued $f \in C_0(\Delta)$. Then show that

$$\phi(f) = \int f \, d\mu,$$

where $\mu = \mu_1 + i\mu_2$.

(b) Let Δ be a second countable, locally compact Hausdorff space, and let $C_0(\Delta)$ denote the space of continuous complex-valued functions on Δ that vanish at infinity. Prove that there is a 1-1 correspondence between the set of all finite complex Borel measures on Δ and the set of all bounded linear functionals on $C_0(\Delta)$.

REMARK. The hypothesis of second countability may be removed from the Riesz Representation Theorem. However, the notion of measurability must be reformulated. Indeed, the σ-algebra on which the measure is defined is, from Theorem 1.1, the smallest σ-algebra for which each element $f \in C_c(\Delta)$ is a measurable function. One can show that this σ-algebra is the smallest σ-algebra containing the compact G_δ sets. This σ-algebra is called the σ-algebra of *Baire* sets, and a measure defined on this σ-algebra is called a *Baire* measure. One can prove versions of Theorems 1.2-1.5, for an arbitrary locally compact Hausdorff space Δ, almost verbatim, only replacing the word "Borel" by the word "Baire."

CHAPTER II

THE HAHN-BANACH EXTENSION THEOREMS AND EXISTENCE OF LINEAR FUNCTIONALS

In this chapter we deal with the problem of extending a linear functional on a subspace Y to a linear functional on the whole space X. The quite abstract results that the Hahn-Banach Theorem comprises (Theorems 2.1, 2.2, 2.3, and 2.6) are, however, of significant importance in analysis, for they provide existence proofs. Applications are made already in this chapter to deduce the existence of remarkable mathematical objects known as Banach limits and translation-invariant measures. One may wish to postpone these applications as well as Theorems 2.4 and 2.5 to a later time. However, the set of exercises concerning convergence of nets should not be omitted, for they will be needed later on.

Let X be a real vector space and let $B = \{x_\alpha\}$, for α in an index set $\mathcal{A}$, be a basis for X. Given any set $\{t_\alpha\}$ of real numbers, also indexed by $\mathcal{A}$, we may define a linear transformation $\phi : X \to \mathbb{R}$ by

$$\phi(x) = \phi\left(\sum c_\alpha x_\alpha\right) = \sum c_\alpha t_\alpha,$$

where $x = \sum c_\alpha x_\alpha$. Note that the sums above are really finite sums, since only finitely many of the coefficients c_α are nonzero for any given x. This ϕ is a linear functional.

EXERCISE 2.1. Prove that if x and y are distinct vectors in a real vector space X, then there exists a linear functional ϕ such that $\phi(x) \neq \phi(y)$. That is, there exist enough linear functionals on X to separate points.

More interesting than the result of the previous exercise is whether there exist linear functionals with some additional properties such as positivity, continuity, or multiplicativity. As we proceed, we will make precise what these additional properties should mean. We begin, motivated by the Riesz representation theorems of the preceding chapter, by studying the existence of positive linear functionals. See Theorem 2.1 below. To do this, we must first make sense of the notion of positivity in a general vector space.

DEFINITION. Let X be a real vector space. By a cone or *positive cone* in X we shall mean a subset P of X satisfying

(1) If x and y are in P, then $x + y$ is in P.

(2) If x is in P and t is a positive real number, then tx is in P.

Given vectors $x_1, x_2 \in X$, we say that $x_1 \geq x_2$ if $x_1 - x_2 \in P$.

Given a positive cone $P \subseteq X$, we say that a linear functional f on X is *positive*, if $f(x) \geq 0$ whenever $x \in P$.

EXERCISE 2.2. (a) Prove that the set of nonnegative functions in a vector space of real-valued functions forms a cone.

(b) Show that the set of nonpositive functions in a vector space of real-valued functions forms a cone.

(c) Let P be the set of points (x, y, z) in $\mathbb{R}^3$ for which $x > \sqrt{y^2 + z^2}$. Prove that P is a cone in the vector space $\mathbb{R}^3$.

THEOREM 2.1. (Hahn-Banach Theorem, Positive Cone Version) *Let P be a cone in a real vector space X, and let Y be a subspace of X having the property that for each $x \in X$ there exists a $y \in Y$ such that $y \geq x$; i.e., $y - x \in P$. Suppose f is a positive linear functional on Y, i.e., $f(y) \geq 0$ if $y \in P \cap Y$. Then there exists a linear functional g on X such that*

(1) *For each $y \in Y$, $g(y) = f(y)$; i.e., g is an extension of f.*

(2) *$g(x) \geq 0$ if $x \in P$; i.e., g is a positive linear functional on X.*

PROOF. Applying the hypotheses both to x and to $-x$, we see that: Given $x \in X$, there exists a $y \in Y$ such that $y - x \in P$, and there exists a $y' \in Y$ such that $y' - (-x) = y' + x \in P$. We will use the existence of these elements of Y later on.

Let S be the set of all pairs (Z, h), where Z is a subspace of X that contains Y, and where h is a positive linear functional on Z that is an extension of f. Since the pair (Y, f) is clearly an element of S, we have that S is nonempty.

Introduce a partial ordering on S by setting

$$(Z, h) \leq (Z', h')$$

if Z is a subspace of Z' and h' is an extension of h, that is $h'(z) = h(z)$
for all $z \in Z$. By the Hausdorff maximality principle, let $\{(Z_\alpha, h_\alpha)\}$ be
a maximal linearly ordered subset of S. Clearly, $Z = \cup Z_\alpha$ is a subspace
of X. Also, if $z \in Z$, then $z \in Z_\alpha$ for some α. Observe that if $z \in Z_\alpha$ and
$z \in Z_\beta$, then, without loss of generality, we may assume that $(Z_\alpha, h_\alpha) \leq$
(Z_β, h_β). Therefore, $h_\alpha(z) = h_\beta(z)$, so that we may uniquely define a
number $h(z) = h_\alpha(z)$, whenever $z \in Z_\alpha$.

We claim that the function h defined above is a linear functional on
the subspace Z. Thus, let z and w be elements of Z. Then $z \in Z_\alpha$ and
$w \in Z_\beta$ for some α and β. Since the set $\{(Z_\gamma, h_\gamma)\}$ is linearly ordered,
we may assume, again without loss of generality, that $Z_\alpha \subseteq Z_\beta$, whence
both z and w are in Z_β. Therefore,

$$h(tz + sw) = h_\beta(tz + sw) = th_\beta(z) + sh_\beta(w) = th(z) + sh(w),$$

showing that h is a linear functional.

Note that, if $y \in Y$, then $h(y) = f(y)$, so that h is an extension of f.
Also, if $z \in Z \cap P$, then $z \in Z_\alpha \cap P$ for some α, whence

$$h(z) = h_\alpha(z) \geq 0,$$

showing that h is a positive linear functional on Z.

We prove next that Z is all of X, and this will complete the proof of
the theorem. Suppose not, and let v be an element of X which is not
in Z. We will derive a contradiction to the maximality of the linearly
ordered subset $\{(Z_\alpha, h_\alpha)\}$ of the partially ordered set S. Let Z' be the
set of all vectors in X of the form $z + tv$, where $z \in Z$ and $t \in \mathbb{R}$. Then
Z' is a subspace of X which properly contains Z.

Let Z_1 be the set of all $z \in Z$ for which $z - v \in P$, and let Z_2 be the
set of all $z' \in Z$ for which $z' + v \in P$. We have seen that both Z_1 and
Z_2 are nonempty. We make the following observation. If $z \in Z_1$ and
$z' \in Z_2$, then $h(z') \geq -h(z)$. Indeed, $z + z' = z - v + z' + v \in P$. So,

$$h(z + z') = h(z) + h(z') \geq 0,$$

and $h(z') \geq -h(z)$, as claimed. Hence, we see that the set B of numbers
$\{h(z')\}$ for which $z' \in Z_2$ is bounded below. In fact, any number of

the form $-h(z)$ for $z \in Z_1$ is a lower bound for B. We write $b = \inf B$. Similarly, the set A of numbers $\{-h(z)\}$ for which $z \in Z_1$ is bounded above, and we write $a = \sup A$. Moreover, we see that $a \leq b$. Note that if $z \in Z_1$, then $h(z) \geq -a$.

Choose any c for which $a \leq c \leq b$, and define h' on Z' by

$$h'(z + tv) = h(z) - tc.$$

Clearly, h' is a linear functional on Z' that extends h and hence extends f. Let us show that h' is a positive linear functional on Z'. On the one hand, if $z + tv \in P$, and if $t > 0$, then $z/t \in Z_2$, and

$$h'(z + tv) = th'((z/t) + v) = t(h(z/t) - c) \geq t(b - c) \geq 0.$$

On the other hand, if $t < 0$ and $z + tv = |t|((z/|t|) - v) \in P$, then $z/(-t) = z/|t| \in Z_1$, and

$$h'(z + tv) = |t|h'((z/(-t)) - v) = |t|(h(z/(-t)) + c) \geq |t|(c - a) \geq 0.$$

Hence, h' is a positive linear functional, and therefore $(Z', h') \in S$. But since $(Z, h) \leq (Z', h')$, it follows that the set $\{(Z_\alpha, h_\alpha)\}$ together with (Z', h') constitutes a strictly larger linearly ordered subset of S, which is a contradiction. Therefore, Z is all of X, h is the desired extension g of f, and the proof is complete.

REMARK. The impact of the Hahn-Banach Theorem is the existence of linear functionals having specified properties. The above version guarantees the existence of many positive linear functionals on a real vector space X, in which there is defined a positive cone. All we need do is find a subspace Y, satisfying the condition in the theorem, and then any positive linear functional on Y has a positive extension to all of X.

EXERCISE 2.3. (a) Verify the details showing that the ordering $\leq$ introduced on the set S in the preceding proof is in fact a partial ordering.

(b) Verify that the function h' defined in the preceding proof is a linear functional on Z'.

(c) Suppose ϕ is a linear functional on the subspace Z' of the above proof. Show that, if ϕ is an extension of h and is a positive linear functional on Z', then the number $\phi(v)$ must be between the numbers a and b of the preceding proof.

EXERCISE 2.4. Let X be a vector space of bounded real-valued functions on a set S. Let P be the cone of nonnegative functions in X. Show that any subspace Y of X that contains the constant functions satisfies the hypothesis of Theorem 2.1.

We now investigate linear functionals that are, in some sense, bounded.

DEFINITION. By a *seminorm* on a real vector space X, we shall mean a real-valued function ρ on X that satisfies:

(1) $\rho(x) \geq 0$ for all $x \in X$,
(2) $\rho(x + y) \leq \rho(x) + \rho(y)$, for all $x, y \in X$, and
(3) $\rho(tx) = |t|\rho(x)$, for all $x \in X$ and all t.

If, in addition, ρ satisfies $\rho(x) = 0$ if and only if $x = 0$, then ρ is called a *norm*, and $\rho(x)$ is frequently denoted by $\|x\|$ or $\|x\|_\rho$. If X is a vector space on which a norm is defined, then X is called a *normed linear space*.

A weaker notion than that of a seminorm is that of a subadditive functional, which is the same as a seminorm except that we drop the nonnegativity condition (condition (1)) and weaken the homogeneity in condition (3). That is, a real-valued function ρ on a real vector space X is called a *subadditive functional* if:

(1) $\rho(x + y) \leq \rho(x) + \rho(y)$ for all $x, y \in X$, and
(2) $\rho(tx) = t\rho(x)$ for all $x \in X$ and $t \geq 0$.

EXERCISE 2.5. Determine whether or not the following are seminorms (subadditive functionals, norms) on the specified vector spaces.

(a) $X = L^p(\mathbb{R})$, $\rho(f) = \|f\|_p = (\int |f|^p)^{1/p}$, for $1 \leq p < \infty$.

(b) X any vector space, $\rho(x) = |f(x)|$, where f is a linear functional on X.

(c) X any vector space, $\rho(x) = \sup_\nu |f_\nu(x)|$, where $\{f_\nu\}$ is a collection of linear functionals on X.

(d) $X = C_0(\Delta)$, $\rho(f) = \|f\|_\infty$, where Δ is a locally compact Hausdorff topological space.

(e) X is the vector space of all infinitely differentiable functions on $\mathbb{R}$, n, m, k are nonnegative integers, and

$$\rho(f) = \sup_{|x| \leq N} \sup_{0 \leq j \leq k} \sup_{0 \leq m \leq M} |x^m f^{(j)}(x)|.$$

(f) X is the set of all bounded real-valued functions on a set S, and $\rho(f) = \sup f(x)$.

(g) X is the space l^∞ of all bounded sequences $\{a_1, a_2, ...\}$, and

$$\rho(\{a_n\}) = \limsup a_n.$$

REMARK. Theorem 2.2 below is perhaps the most familiar version of the Hahn-Banach theorem. So, although it can be derived as a consequence of Theorem 2.1 and is in fact equivalent to that theorem (see parts d and e of Exercise 2.6), we give here an independent proof.

THEOREM 2.2. (Hahn-Banach Theorem, Seminorm Version) *Let ρ be a seminorm on a real vector space X. Let Y be a subspace of X, let f be a linear functional on Y, and assume that*

$$f(y) \leq \rho(y)$$

for all $y \in Y$. Then there exists a linear functional g on X, which is an extension of f and which satisfies

$$g(x) \leq \rho(x)$$

for all $x \in X$.

PROOF. By analogy with the proof of Theorem 2.1, we let S be the set of all pairs (Z, h), where Z is a subspace of X containing Y, h is a linear functional on Z that extends f, and $h(z) \leq \rho(z)$ for all $z \in Z$. We give to S the same partial ordering as in the preceding proof. By the Hausdorff maximality principle, let $\{(Z_\alpha, h_\alpha)\}$ be a maximal linearly ordered subset of S. As before, we define $Z = \cup Z_\alpha$, and h on Z by $h(z) = h_\alpha(z)$ whenever $z \in Z_\alpha$. It follows as before that h is a linear functional on Z, that extends f, for which $h(z) \leq \rho(z)$ for all $z \in Z$, so that the proof will be complete if we show that $Z = X$.

Suppose that $Z \neq X$, and let v be a vector in X which is not in Z. Define Z' to be the set of all vectors of the form $z + tv$, for $z \in Z$ and $t \in \mathbb{R}$. We observe that for any z and z' in Z,

$$h(z) + h(z') = h(z + z') \leq \rho(z + v + z' - v) \leq \rho(z + v) + \rho(z' - v),$$

or that

$$h(z') - \rho(z' - v) \leq \rho(z + v) - h(z).$$

Let A be the set of numbers $\{h(z') - \rho(z' - v)\}$ for $z' \in Z$, and put $a = \sup A$. Let B be the numbers $\{\rho(z + v) - h(z)\}$ for $z \in Z$, and put $b = \inf B$. It follows from the calculation above that $a \leq b$. Choose c to be any number for which $a \leq c \leq b$, and define h' on Z' by

$$h'(z + tv) = h(z) + tc.$$

Obviously h' is linear and extends f. If $t > 0$, then

$$\begin{aligned}
h'(z + tv) &= t(h(z/t) + c) \\
&\leq t(h(z/t) + b) \\
&\leq t(h(z/t) + \rho((z/t) + v) - h(z/t)) \\
&= t\rho((z/t) + v) \\
&= \rho(z + tv).
\end{aligned}$$

And, if $t < 0$, then

$$\begin{aligned}
h'(z + tv) &= |t|(h(z/|t|) - c) \\
&\leq |t|(h(z/|t|) - a) \\
&\leq |t|(h(z/|t|) - h(z/|t|) + \rho((z/|t|) - v)) \\
&= |t|\rho((z/|t|) - v) \\
&= \rho(z + tv),
\end{aligned}$$

which proves that $h'(z + tv) \leq \rho(z + tv)$ for all $z + tv \in Z'$.

Hence, $(Z', h') \in S$, $(Z, h) < (Z', h')$, and the maximality of the linearly ordered set $\{(Z_\alpha, h_\alpha)\}$ is contradicted. This completes the proof.

THEOREM 2.3. (Hahn-Banach Theorem, Norm Version) *Let Y be a subspace of a normed linear space X, and suppose that f is a linear functional on Y for which there exists a positive constant M satisfying $|f(y)| \leq M\|y\|$ for all $y \in Y$. Then there exists an extension of f to a linear functional g on X satisfying $|g(x)| \leq M\|x\|$ for all $x \in X$.*

EXERCISE 2.6. (a) Prove the preceding theorem.

(b) Let the notation be as in the proof of Theorem 2.2. Suppose ϕ is a linear functional on Z' that extends the linear functional h and for which $\phi(z') \leq \rho(z')$ for all $z' \in Z'$. Prove that $\phi(v)$ must satisfy $a \leq \phi(v) \leq b$.

(c) Show that Theorem 2.2 holds if the seminorm ρ is replaced by the weaker notion of a subadditive functional.

(d) Derive Theorem 2.2 as a consequence of Theorem 2.1. HINT: Let $X' = X \oplus \mathbb{R}$, Define P to be the set of all $(x, t) \in X'$ for which $\rho(x) \leq t$, let $Y' = Y \oplus \mathbb{R}$, and define f' on Y' by $f'(y, t) = t - f(y)$. Now apply Theorem 2.1.

(e) Derive Theorem 2.1 as a consequence of Theorem 2.2. HINT: Define ρ on X by $\rho(x) = \inf f(y)$, where the infimum is taken over all $y \in Y$ for which $y - x \in P$. Show that ρ is a seminorm, and then apply Theorem 2.2.

We devote the next few exercises to developing the notion of convergence of nets. This topological concept is of great use in functional analysis. The reader should notice how crucial the axiom of choice is in these exercises. Indeed, the Tychonoff theorem (Exercise 2.11) is known to be equivalent to the axiom of choice.

DEFINITION. A *directed set* is a nonempty set D, on which there is defined a transitive and reflexive partial ordering $\leq$, satisfying the following condition: If $\alpha, \beta \in D$, then there exists an element $\gamma \in D$ such that $\alpha \leq \gamma$ and $\beta \leq \gamma$. That is, every pair of elements of D has an upper bound.

If C and D are two directed sets, and h is a mapping from C into D, then h is called *order-preserving* if $c_1 \leq c_2$ implies that $h(c_1) \leq h(c_2)$. An order-preserving map h of C into D is called *cofinal* if for each $\alpha \in D$ there exists a $\beta \in C$ such that $\alpha \leq h(\beta)$.

A *net* in a set X is a function f from a directed set D into X. A net f in X is frequently denoted, in analogy with a sequence, by $\{x_\alpha\}$, where $x_\alpha = f(\alpha)$.

If $\{x_\alpha\}$ denotes a net in a set X, then a *subnet* of $\{x_\alpha\}$ is determined by an order-preserving cofinal function h from a directed set C into D, and is the net g defined on C by $g(\beta) = x_{h(\beta)}$. The values $h(\beta)$ of the function h are ordinarily denoted by $h(\beta) = \alpha_\beta$, whence the subnet g takes the notation $g(\beta) = x_{\alpha_\beta}$.

A net $\{x_\alpha\}$, $\alpha \in D$, in a topological space X is said to *converge* to an element $x \in X$, and we write $x = \lim_\alpha x_\alpha$, if For each open set U containing x, there exists an $\alpha \in D$ such that $x'_\alpha \in U$ whenever $\alpha \leq \alpha'$.

EXERCISE 2.7. (a) Show that any linearly ordered set is a directed set.

(b) Let S be a set and let D be the set of all finite subsets F of S. Show that D is a directed set if the partial ordering on D is given by $F_1 \leq F_2$ if and only if $F_1 \subseteq F_2$.

(c) Let x be a point in a topological space X, and let D be the partially-ordered set of all neighborhoods of x with the ordering $U \leq V$ if and only if $V \subseteq U$. Prove that D is a directed set.

(d) Let D and D' be directed sets. Show that $D \times D'$ is a directed set, where the ordering is given by $(\alpha, \alpha') \leq (\beta, \beta')$ if and only if $\alpha \leq \alpha'$ and $\beta \leq \beta'$.

(e) Verify that every sequence is a net.

(f) Let $\{x_n\}$ be a sequence. Show that there exist subnets of the net $\{x_n\}$ which are not subsequences.

EXERCISE 2.8. (a) (Uniqueness of Limits) Let $\{x_\alpha\}$ be a net in a Hausdorff topological space X. Suppose $x = \lim x_\alpha$ and $y = \lim x_\alpha$. Show that $x = y$.

(b) Suppose $\{x_\alpha\}$ and $\{y_\alpha\}$ are nets (defined on the same directed set D) in $\mathbb{C}$, and assume that $x = \lim x_\alpha$ and $y = \lim y_\alpha$. Prove that

$$x + y = \lim(x_\alpha + y_\alpha),$$

$$xy = \lim(x_\alpha y_\alpha),$$

and that if $a \leq x_\alpha \leq b$ for all α, then

$$a \leq x \leq b.$$

(c) Prove that if a net $\{x_\alpha\}$ converges to an element x in a topological space X, then every subnet $\{x_{\alpha_\beta}\}$ of $\{x_\alpha\}$ also converges to x.

(d) Prove that a net $\{x_\alpha\}$ in a topological space X converges to an element $x \in X$ if and only if every subnet $\{x_{\alpha_\beta}\}$ of $\{x_\alpha\}$ has in turn a subnet $\{x_{\alpha_{\beta_\gamma}}\}$ that converges to x. HINT: To prove the "if" part, argue by contradiction. Use the directed set $\mathcal{U} \times D$, where $\mathcal{U}$ is the set of all neighborhoods of x.

(e) Let A be a subset of a topological space X. We say that an element $x \in X$ is a *cluster point* of A if there exists a net $\{x_\alpha\}$ in A such that $x = \lim x_\alpha$. Prove that A is closed if and only if it contains all of its cluster points.

(f) Let f be a function from a topological space X into a topological space Y. Show that f is continuous at a point $x \in X$ if and only if for each net $\{x_\alpha\}$ that converges to $x \in X$, the net $\{f(x_\alpha)\}$ converges to $f(x) \in Y$.

EXERCISE 2.9. (a) Let X be a compact topological space. Show that every net in X has a convergent subnet. HINT: Let $\{x_\alpha\}$ be a net in X defined on a directed set D. For each $\alpha \in D$, define $V_\alpha \subseteq X$ to be the set of all $x \in X$ for which there exists a neighborhood U_x of X such that $x_\beta \notin U_x$ whenever $\alpha \leq \beta$. Show that, if $x \notin \cup V_\alpha$, then x is the limit of some subnet of $\{x_\alpha\}$. Now, argue by contradiction.

(b) Prove that a topological space X is compact if and only if every net in X has a convergent subnet. HINT: Let $\mathcal{F}$ be a collection of closed subsets of X for which the intersection of any finite number of elements of $\mathcal{F}$ is nonempty. Let D be the directed set whose elements are the finite subsets of $\mathcal{F}$.

(c) Let $\{x_\alpha\}$ be a net in a metric space X. Define what it means for the net $\{x_\alpha\}$ to be a *Cauchy* net. Show that, if X is a complete metric space, then a net $\{x_\alpha\}$ is convergent if and only if it is a Cauchy net.

EXERCISE 2.10. Let X be a set, let $\{f_i\}$, for i in an index set I, be a collection of real-valued functions on X, and let $\mathcal{T}$ be the weakest topology on X for which each f_i is continuous.

(a) Show that a net $\{x_\alpha\}$ in the topological space $(X, \mathcal{T})$ converges to an element $x \in X$ if and only if

$$f_i(x) = \lim_\alpha f_i(x_\alpha)$$

for every $i \in I$.

(b) Let X be a set, for each $x \in X$ let Y_x be a topological space, and let Y be the topological product space

$$Y = \prod_{x \in X} Y_x.$$

Prove that a net $\{y_\alpha\}$ in Y converges if and only if, for each $x \in X$, the net $\{y_\alpha(x)\}$ converges in Y_x.

EXERCISE 2.11. Prove the Tychonoff Theorem. That is, prove that if $X = \prod_{i \in I} X_i$, where each X_i is a compact topological space, then X is a compact topological space. HINT: Let $\{x_\alpha\}$ be a net in X, defined on a directed set D. Show that there exists a convergent subnet as follows:

(a) Prove by induction on n that, for each finite subset $F \subseteq I$ of cardinality n, there exists a directed set D_F and an order-preserving cofinal map $h_F : D_F \to D$ for which the subnet $\{x_{h_F(\beta)}\}$ satisfies:

(1) The net $\{(x_{h_F(\beta)})_i\}$ converges in X_i for each $i \in F$.
(2) If F' is a subset of F, then the net $\{x_{h_F(\beta)}\}$ is a subnet of the net $\{x_{h_{F'}(\gamma)}\}$. That is, there exists an order-preserving cofinal map $h_{F',F} : D_F \to D_{F'}$ satisfying $h_{F'}(h_{F',F}(\beta)) = h_F(\beta)$ for all $\beta \in D_F$.

(b) Let $\mathcal{F}$ denote the set of all finite subsets of I, and let C be the set of all pairs (F, β) for $F \in \mathcal{F}$ and $\beta \in D_F$. Define a partial ordering on C by $(F', \gamma) \leq (F, \beta)$ if and only if $F' \subseteq F$ and $\gamma \leq h_{F',F}(\beta)$. Define a map $h : C \to D$ by $h(F, \beta) = h_F(\beta)$. Prove that C is a directed set, that h is an order-preserving cofinal map of C into D, and that the subnet $\{x_{h(F,\beta)}\}$ of $\{x_\alpha\}$ converges in X.

EXERCISE 2.12. (a) Suppose $\{f_\alpha\}$ is a net of linear functionals on a vector space X, and suppose that the net converges pointwise to a function f. Prove that f is a linear functional.

(b) Suppose ρ is a subadditive functional on a vector space X and that $x \in X$. Prove that $-\rho(-x) \leq \rho(x)$.

(c) Suppose ρ is a subadditive functional on a vector space X, and let F^ρ be the set of all linear functionals f on X for which $f(x) \leq \rho(x)$ for every $x \in X$. Let K be the compact Hausdorff space

$$K = \prod_{x \in X} [-\rho(-x), \rho(x)]$$

(thought of as a space of functions on X). Prove that F^ρ is a closed subset of K. Conclude that F^ρ is a compact Hausdorff space in the topology of pointwise convergence on X.

THEOREM 2.4. *Let ρ be a subadditive functional on a vector space X, and let g be a linear functional on X such that $g(x) \leq \rho(x)$ for all $x \in X$. Suppose γ is a linear transformation of X into itself for which $\rho(\gamma(x)) = \rho(x)$ for all $x \in X$. Then there exists a linear functional h on X satisfying:*

(1) $h(x) \leq \rho(x)$ *for all $x \in X$.*
(2) $h(\gamma(x)) = h(x)$ *for all $x \in X$.*
(3) *If $x \in X$ satisfies $g(x) = g(\gamma^n(x))$ for all positive n, then $h(x) = g(x)$.*

PROOF. For each positive integer n, define

$$g_n(x) = (1/n) \sum_{i=1}^{n} g(\gamma^i(x)).$$

Let F^ρ and K be as in the preceding exercise. Then the sequence $\{g_n\}$ is a net in the compact Hausdorff space K, and consequently there exists a convergent subnet $\{g_{n_\alpha}\}$. By Exercise 2.12, we know then that the subnet $\{g_{n_\alpha}\}$ of the sequence (net) $\{g_n\}$ converges pointwise to a linear functional h on X and that $h(x) \leq \rho(x)$ for all $x \in X$.

Using the fact that $-\rho(x) \leq g(\gamma^i(x)) \leq \rho(x)$ for all $x \in X$ and all $i > 0$, and the fact that the cofinal map $\alpha \to n_\alpha$ diverges to infinity, we have that

$$h(\gamma(x)) = \lim_{\alpha} g_{n_\alpha}(x)$$

$$= \lim_{\alpha}(1/n_\alpha) \sum_{i=1}^{n_\alpha} g(\gamma^{i+1}(x))$$

$$= \lim_{\alpha}(1/n_\alpha) \sum_{i=2}^{n_\alpha+1} g(\gamma^i(x))$$

$$= \lim_{\alpha}(1/n_\alpha)[\sum_{i=1}^{n_\alpha} g(\gamma^i(x)) + g(\gamma^{n_\alpha+1}(x)) - g(\gamma(x))]$$

$$= \lim_{\alpha}(1/n_\alpha) \sum_{i=1}^{n_\alpha} g(\gamma^i(x))$$

$$= \lim_{\alpha} g_{n_\alpha}(x)$$

$$= h(x),$$

which proves the second statement of the theorem.

Finally, if x is such that $g(\gamma^n(x)) = g(x)$ for all positive n, then $g_n(x) = g(x)$ for all n, whence $h(x) = g(x)$, and this completes the proof.

EXERCISE 2.13. (Banach Means) Let $X = l^\infty$ be the vector space of all bounded sequences $\{a_1, a_2, a_3, \ldots\}$ of real numbers. A *Banach mean* or *Banach limit* is a linear functional M on X such that for all $\{a_n\} \in X$ we have:

$$\inf a_n \leq M(\{a_n\}) \leq \sup a_n.$$

and

$$M(\{a_{n+1}\}) = M(\{a_n\}).$$

(a) Prove that there exists a Banach limit on X. HINT: Use Theorem 2.2, or more precisely part c of Exercise 2.6, with Y the subspace of constant sequences, f the linear functional sending a constant sequence to that constant, and ρ the subadditive functional given by $\rho(\{a_n\}) = \limsup a_n$. Then use Theorem 2.4 applied to the extension g of f. (Note that, since the proof to Theorem 2.4 depends on the Tychonoff theorem, the very existence of Banach means depends on the axiom of choice.)

(b) Show that any Banach limit M satisfies $M(\{a_n\}) = L$, if $L = \lim a_n$, showing that any Banach limit is a generalization of the ordinary notion of limit.

(c) Show that any Banach limit assigns the number $1/2$ to the sequence $\{0, 1, 0, 1, \ldots\}$.

(d) Construct a sequence $\{b_n\} \in X$ which does not converge but for which

$$\lim(b_{n+1} - b_n) = 0.$$

Show that any linear functional g on X, for which $g(\{a_n\}) \leq \limsup a_n$ for all $\{a_n\} \in X$, satisfies $g(\{b_n\}) = g(\{b_{n+1}\})$.

(e) Use the sequence $\{b_n\}$ of part d to prove that there exist uncountably many distinct Banach limits on X. HINT: Use the Hahn-Banach Theorem and Theorem 2.4 to find a Banach limit that takes the value r on this sequence, where r is any number satisfying $\liminf b_n \leq r \leq \limsup b_n$.

EXERCISE 2.14. Prove the following generalization of Theorem 2.4. Let ρ be a subadditive functional on a vector space X, and let g be a linear functional on X such that $g(x) \leq \rho(x)$ for all $x \in X$. Suppose $\gamma_1, \ldots, \gamma_n$ are commuting linear transformations of X into itself for which $\rho(\gamma_i(x)) = \rho(x)$ for all $x \in X$ and all $1 \leq i \leq n$. Then there exists a linear functional h on X satisfying:

(1) $h(x) \leq \rho(x)$ for all $x \in X$.

(2) $h(\gamma_i(x)) = h(x)$ for all $x \in X$ and all $1 \leq i \leq n$.

(3) If $x \in X$ satisfies $g(x) = g(\gamma_i^k(x))$ for all positive k and all $1 \leq i \leq n$, then $h(x) = g(x)$.

HINT: Use Theorem 2.4 and mathematical induction.

THEOREM 2.5. (Hahn-Banach Theorem, Semigroup-Invariant Version) *Let ρ be a subadditive functional on a real vector space X, and let f be a linear functional on a subspace Y of X for which $f(y) \leq \rho(y)$ for all $y \in Y$. Suppose Γ is an abelian semigroup of linear transformations of X into itself for which:*

(1) $\rho(\gamma(x)) = \rho(x)$ *for all $\gamma \in \Gamma$ and $x \in X$; i.e., ρ is invariant under Γ.*

(2) $\gamma(Y) \subseteq Y$ *for all $\gamma \in \Gamma$; i.e., Y is invariant under Γ.*

(3) $f(\gamma(y)) = f(y)$ *for all $\gamma \in \Gamma$ and $y \in Y$; i.e., f is invariant under Γ.*

Then there exists a linear functional g on X for which

(a) *g is an extension of f.*

(b) *$g(x) \leq \rho(x)$ for all $x \in X$.*

(c) *$g(\gamma(x)) = g(x)$ for all $\gamma \in \Gamma$ and $x \in X$; i.e., g is invariant under Γ.*

PROOF. For A a finite subset of Γ, we use part c of Exercise 2.6 and then Exercise 2.14 to construct a linear functional g_A on X satisfying:

(1) g_A is an extension of f.
(2) $g_A(x) \leq \rho(x)$ for all $x \in X$.
(3) $g_A(\gamma(x)) = g_A(x)$ for all $x \in X$ and $\gamma \in A$.

If as in Exercise 2.12 $K = \prod_{x \in X}[-\rho(-x), \rho(x)]$, then $\{g_A\}$ can be regarded as a net in the compact Hausdorff space K. Let $\{g_{A_\beta}\}$ be a convergent subnet, and write $h = \lim_\beta g_{A_\beta}$. Then, h is a function on X, and is in fact the pointwise limit of a net of linear functionals, and so is itself a linear functional.

Clearly, $h(x) \leq \rho(x)$ for all $x \in X$, and h is an extension of f.

To see that $h(\gamma(x)) = h(x)$ for all $\gamma \in \Gamma$, fix a γ_0, and let $A_0 = \{\gamma_0\}$. By the definition of a subnet, there exists a β_0 such that if $\beta \geq \beta_0$ then $A_\beta \geq A_0$. Hence, if $\beta \geq \beta_0$, then $\{\gamma_0\} \subseteq A_\beta$. So, if $\beta \geq \beta_0$, then $g_{A_\beta}(\gamma_0(x)) = g_{A_\beta}(x)$ for all x. Hence,

$$h(\gamma_0(x)) = \lim_\beta g_{A_\beta}(\gamma_0(x)) = \lim_\beta g_{A_\beta}(x) = h(x),$$

as desired.

DEFINITION. Let S be a set. A *ring* of subsets of S is a collection $\mathcal{R}$ of subsets of S such that if $E, F \in \mathcal{R}$, then both $E \cup F$ and $E \Delta F$ are in $\mathcal{R}$, where $E \Delta F = (E \cap \tilde{F}) \cup (F \cap \tilde{E})$ is the symmetric difference of E and F. By a *finitely additive measure* on S, we mean an assignment $E \to \mu(E)$, of a ring $\mathcal{R}$ of subsets of S into the extended nonnegative real numbers, such that

$$\mu(\emptyset) = 0$$

and

$$\mu(E_1 \cup \ldots \cup E_n) = \mu(E_1) + \ldots + \mu(E_n)$$

whenever $\{E_1 \ldots, E_n\}$ is a pairwise disjoint collection of elements of $\mathcal{R}$.

EXERCISE 2.15. (Translation-Invariant Finitely Additive Measures) Let X be the vector space of all bounded functions on $\mathbb{R}$ with compact support, and let P be the positive cone of nonnegative functions in X.

(a) Let I be a positive linear functional on X. For each bounded subset $E \subset \mathbb{R}$, define $\mu(E) = I(\chi_E)$. Show that the set of all bounded subsets of $\mathbb{R}$ is a ring $\mathcal{R}$ of sets and that μ is a finitely additive measure on this ring.

(b) Show that there exists a finitely additive measure ν, defined on the ring of all bounded subsets of $\mathbb{R}$, such that $\nu(E)$ is the Lebesgue measure for every bounded Lebesgue measurable subset E of $\mathbb{R}$, and such that $\nu(E+x) = \nu(E)$ for all bounded subsets E of $\mathbb{R}$ and all real numbers x. (Such a measure is said to be *translation-invariant*.) HINT: Let Y be the subspace of X consisting of the bounded Lebesgue measurable functions of bounded support, let $I(f) = \int f$, and let Γ be the semigroup of linear transformations of X determined by the semigroup of all translations of $\mathbb{R}$. Now use Theorem 2.5.

(c) Let μ be the finitely additive measure of part b. For each subset E of $\mathbb{R}$, define $\nu(E) = \lim_n \mu(E \cap [-n, n])$. Prove that ν is a translation-invariant, finitely additive measure on the σ-algebra of all subsets of $\mathbb{R}$, and that μ agrees with Lebesgue measure on Lebesgue measurable sets.

(d) Prove that there exists no countably additive translation-invariant measure μ on the σ-algebra of all subsets of $\mathbb{R}$ that agrees with Lebesgue measure on Lebesgue measurable sets. HINT: Suppose μ is such a countably additive measure. Define an equivalence relation on $\mathbb{R}$ by setting $x \equiv y$ if $y - x \in \mathbb{Q}$, i.e., $y - x$ is a rational number. Let $E \subset (0, 1)$ be a set of representatives of the equivalence classes of this relation. Show first that $\cup_{q \in \mathbb{Q} \cap (0,1)} E + q \subset (0, 2)$, whence $\mu(E)$ must be 0. Then show that $(0, 1) \subset \cup_{q \in \mathbb{Q}} E + q$, whence $\mu(E)$ must be positive.

DEFINITION. Let X be a complex vector space. A *seminorm* on X is a real-valued function ρ that is *subadditive* and *absolutely homogeneous*; i.e.,

$$\rho(x + y) \leq \rho(x) + \rho(y)$$

for all $x, y \in X$, and

$$\rho(\lambda x) = |\lambda| \rho(x)$$

for all $x \in X$ and $\lambda \in \mathbb{C}$.

THEOREM 2.6. (Hahn-Banach Theorem, Complex Version) *Let ρ be a seminorm on a complex vector space X. Let Y be a subspace of X, and let f be a complex-linear functional on Y satisfying $|f(y)| \leq \rho(y)$ for all $y \in Y$. Then there exists a complex-linear functional g on X satisfying g is an extension of f, and $|g(x)| \leq \rho(x)$ for all $x \in X$.*

EXERCISE 2.16. Prove Theorem 2.6 as follows:

(a) Use Theorem 2.2 to extend the real part u of f to a real linear functional a on X that satisfies $a(x) \leq \rho(x)$ for all $x \in X$.

(b) Use Exercise 1.11 and part a to define a complex linear functional g on X that extends f.

(c) For $x \in X$, choose a complex number λ of absolute value 1 such that $|g(x)| = \lambda g(x)$. Then show that

$$|g(x)| = g(\lambda x) = a(\lambda x) \leq \rho(x).$$

CHAPTER III

TOPOLOGICAL VECTOR SPACES AND
CONTINUOUS LINEAR FUNCTIONALS

The marvelous interaction between linearity and topology is introduced
in this chapter. Although the most familiar examples of this interaction
may be normed linear spaces, we have in mind here the more subtle, and
perhaps more important, topological vector spaces whose topologies are
defined as the weakest topologies making certain collections of functions
continuous. See the examples in Exercises 3.8 and 3.9, and particularly
the Schwartz space $\mathcal{S}$ discussed in Exercise 3.10.

DEFINITION. A *topological vector space* is a real (or complex) vector space X on which there is a Hausdorff topology such that:

(1) The map $(x, y) \to x + y$ is continuous from $X \times X$ into X. (Addition
is continuous.) and

(2) The map $(t, x) \to tx$ is continuous from $\mathbb{R} \times X$ into X (or $\mathbb{C} \times X$
into X). (Scalar multiplication is continuous.)

We say that a topological vector space X is a *real* or *complex* topological vector space according to which field of scalars we are considering.
A complex topological vector space is obviously also a real topological
vector space.

A metric d on a vector space X is called *translation-invariant* if $d(x +
z, y + z) = d(x, y)$ for all $x, y, z \in X$. If the topology on a topological
vector space X is determined by a translation-invariant metric d, we
call X (or (X, d)) a *metrizable* vector space. If x is an element of a
metrizable vector space (X, d), we denote by $B_\epsilon(x)$ the ball of radius
ϵ around x; i.e., $B_\epsilon(x) = \{y : d(x, y) < \epsilon\}$. If the topology on a vector

space X is determined by the translation-invariant metric d defined by a norm on X, i.e., $d(x, y) = \|x - y\|$, we call X a *normable* vector space. If the topology on X is determined by some complete translation-invariant metric, we call X a *Frechet* space.

The topological vector space X is called *separable* if it contains a countable dense subset.

Two topological vector spaces X_1 and X_2 are *topologically isomorphic* if there exists a linear isomorphism T from X_1 onto X_2 that is also a homeomorphism. In this case, T is called a *topological isomorphism.*

EXERCISE 3.1. (a) Let X be a topological vector space, and let x be a nonzero element of X. Show that the map $y \to x + y$ is a (nonlinear) homeomorphism of X onto itself. Hence, U is a neighborhood of 0 if and only if $x + U$ is a neighborhood of x. Show further that if U is an open subset of X and S is any subset of X, then $S + U$ is an open subset of X.

(b) Show that $x \to -x$ is a topological isomorphism of X onto itself. Hence, if U is a neighborhood of 0, then $-U$ also is a neighborhood of 0, and hence $V = U \cap (-U)$ is a *symmetric* neighborhood of 0; i.e., $x \in V$ if and only if $-x \in V$.

(c) If U is a neighborhood of 0 in a topological vector space X, use the continuity of addition to show that there exists a neighborhood V of 0 such that $V + V \subseteq U$.

(d) If $X_1, \ldots, X_n$ are topological vector spaces, show that the (algebraic) direct sum $\bigoplus_{i=1}^{n} X_i$ is a topological vector space, with respect to the product topology. What about the direct product of infinitely many topological vector spaces?

(e) If Y is a linear subspace of X, show that Y is a topological vector space with respect to the relative topology.

(f) Show that, with respect to its Euclidean topology, $\mathbb{R}^n$ is a real topological vector space, and $\mathbb{C}^n$ is a complex topological vector space.

THEOREM 3.1. *Let X be a topological vector space. Then:*

(1) X *is a regular topological space; i.e., if A is a closed subset of X and x is an element of X that is not in A, then there exist disjoint open sets U_1 and U_2 such that $x \in U_1$ and $A \subseteq U_2$.*

(2) X *is connected.*

(3) X *is compact if and only if X is $\{0\}$.*

(4) *Every finite dimensional subspace Y of X is a closed subset of X.*

(5) *If T is a linear transformation of X into another topological*

vector space X', then T is continuous at each point of X if and only if T is continuous at the point $0 \in X$.

PROOF. To see 1, let A be a closed subset of X and let x be a point of X not in A. Let U denote the open set $\tilde{A}$, and let U' be the open neighborhood $U - x$ of 0. (See part a of Exercise 3.1.) Let V be a neighborhood of 0 such that $V + V \subset U'$. Now $-V$ is a neighborhood of 0, and we let $W = V \cap (-V)$. Then $W = -W$ and $W + W \subset U'$. Let $U_1 = W + x$ and let $U_2 = W + A$. Then $x \in U_1$ and $A \subseteq U_2$. Clearly U_1 is an open set, and, because $U_2 = \cup_{y \in A}(W + y)$, we see also that U_2 is an open set. Further, if $z \in U_1 \cap U_2$, then we must have $z = x + w_1$ and $z = a + w_2$, where both w_1 and w_2 belong to W and $a \in A$. But then we would have

$$a = x + w_1 - w_2 \in x + W - W \subset x + U' = U = \tilde{A},$$

which is a contradiction. Therefore, $U_1 \cap U_2 = \emptyset$, and X is a regular topological space.

Because the map $t \to (1 - t)x + ty$ is continuous on $\mathbb{R}$, it follows that any two elements of X can be joined by a curve, in fact by a line segment in X. Therefore, X is pathwise connected, hence connected, proving part 2.

Part 3 is left to an exercise.

We prove part 4 by induction on the dimension of the subspace Y. Although the assertion in part 4 seems simple enough, it is surprisingly difficult to prove. First, if Y has dimension 1, let $y \neq 0 \in Y$ be a basis for Y. If $\{t_\alpha y\}$ is a net in Y that converges to an element $x \in X$, then the net $\{t_\alpha\}$ must be bounded in $\mathbb{R}$ (or $\mathbb{C}$). Indeed, if the net $\{t_\alpha\}$ were not bounded, let $\{t_{\alpha_\beta}\}$ be a subnet for which $\lim_\beta |t_{\alpha_\beta}| = \infty$. Then

$$\begin{aligned} y &= \lim_\beta (1/t_{\alpha_\beta}) t_{\alpha_\beta} y \\ &= \lim_\beta (1/t_{\alpha_\beta}) \lim_\beta t_{\alpha_\beta} y \\ &= 0 \times x \\ &= 0, \end{aligned}$$

which is a contradiction. So, the net $\{t_\alpha\}$ is bounded. Let $\{t_{\alpha_\beta}\}$ be a convergent subnet of $\{t_\alpha\}$ with limit t. Then

$$x = \lim_\alpha t_\alpha y = \lim_\beta t_{\alpha_\beta} y = ty,$$

whence $x \in Y$, and Y is closed.

Assume now that every $n-1$-dimensional subspace is closed, and let Y have dimension $n > 1$. Let $\{y_1, \ldots, y_n\}$ be a basis for Y, and write Y' for the linear span of $y_1, \ldots, y_{n-1}$. Then elements y of Y can be written uniquely in the form $y = y' + ty_n$, for $y' \in Y'$ and t real (complex). Suppose that x is an element of the closure of Y, i.e., $x = \lim_\alpha y'_\alpha + t_\alpha y_n$. As before, we have that the net $\{t_\alpha\}$ must be bounded. Indeed, if the net $\{t_\alpha\}$ were not bounded, then let $\{t_{\alpha_\beta}\}$ be a subnet for which $\lim_\beta |t_{\alpha_\beta}| = \infty$. Then

$$0 = \lim_\beta (1/t_{\alpha_\beta})x = \lim_\beta (y'_{\alpha_\beta}/t_{\alpha_\beta}) + y_n,$$

or

$$y_n = \lim_\beta -(y'_{\alpha_\beta}/t_{\alpha_\beta}),$$

implying that y_n belongs to the closure of the closed subspace Y'. Since y_n is linearly independent of the subspace Y', this is impossible, showing that the sequence $\{t_\alpha\}$ is bounded. Hence, letting $\{t_{\alpha_\beta}\}$ be a convergent subnet of $\{t_\alpha\}$, say $t = \lim_\beta t_{\alpha_\beta}$, we have

$$x = \lim_\beta y'_{\alpha_\beta} + t_{\alpha_\beta} y_n,$$

showing that

$$x - ty_n = \lim y_{\alpha_\beta},$$

whence, since Y' is closed, there exists a $y' \in Y'$ such that $x - ty_n = y'$. Therefore, $x = y' + ty_n \in Y$, and Y is closed, proving part 4.

Finally, if T is a linear transformation from X into X', then T being continuous at every point of X certainly implies that T is continuous at 0. Conversely, suppose T is continuous at 0, and let $x \in X$ be given. If V is a neighborhood of $T(x) \in X'$, let U be the neighborhood $V - T(x)$ of $0 \in X'$. Because T is continuous at 0, there exists a neighborhood W of $0 \in X$ such that $T(W) \subseteq U$. But then the neighborhood $W + x$ of x satisfies $T(W + x) \subseteq U + T(x) = V$, and this shows the continuity of T at x.

EXERCISE 3.2. (a) Prove part 3 of the preceding theorem.

(b) Prove that any linear transformation T, from $\mathbb{R}^n$ (or $\mathbb{C}^n$), equipped with its ordinary Euclidean topology, into a real (complex) topological vector space X, is necessarily continuous. HINT: Let $e_1, \ldots, e_n$ be the standard basis, and write $x_i = T(e_i)$.

(c) Let ρ be a seminorm (or subadditive functional) on a real topological vector space X. Show that ρ is continuous everywhere on X if and only if it is continuous at 0.

(d) Suppose ρ is a continuous seminorm on a real vector space X and that f is a linear functional on X that is bounded by ρ; i.e., $f(x) \leq \rho(x)$ for all $x \in X$. Prove that f is continuous.

(e) Suppose X is a vector space on which there is a topology $\mathcal{T}$ such that $(x, y) \to x + y$ is continuous from $X \times X$ into X. Show that $\mathcal{T}$ is Hausdorff if and only if it is T_0. (A topological space is called T_0 if, given any two points, there exists an open set that contains one of them but not the other.)

(f) Show that $L^p(\mathbb{R})$ is a topological vector space with respect to the topology defined by the (translation-invariant) metric

$$d(f, g) = \|f - g\|_p.$$

Show, in fact, that any normed linear space is a topological vector space with respect to the topology defined by the metric given by

$$d(x, y) = \|x - y\|.$$

(g) Let c_c denote the set of all real (or complex) sequences $\{a_1, a_2, \dots\}$ that are nonzero for only finitely many terms. If $\{a_j\} \in c_c$, define the norm of $\{a_j\}$ by $\|\{a_j\}\| = \max_j |a_j|$. Verify that c_c is a normed linear space with respect to this definition of norm.

(h) Give an example of a (necessarily infinite dimensional) subspace of $L^p(\mathbb{R})$ which is not closed.

THEOREM 3.2. (Finite-Dimensional Topological Vector Spaces)

(1) If X is a finite dimensional real (or complex) topological vector space, and if $x_1, \dots, x_n$ is a basis for X, then the map $T : \mathbb{R}^n \to X$ (or $T : \mathbb{C}^n \to X$), defined by $T(t_1, \dots, t_n) = \sum t_i x_i$, is a topological isomorphism of $\mathbb{R}^n$ (or $\mathbb{C}^n$), equipped with its Euclidean topology, onto X. That is, specifically, a net $\{x_\alpha\} = \{\sum_{i=1}^n t_i^\alpha x_i\}$ converges to an element $x = \sum_{i=1}^n t_i x_i \in X$ if and only if each net $\{t_i^\alpha\}$ converges to t_i, $1 \leq i \leq n$.

(2) The only topology on $\mathbb{R}^n$ (or $\mathbb{C}^n$), in which it is a topological vector space, is the usual Euclidean topology.

(3) Any linear transformation, from one finite dimensional topological vector space into another finite dimensional topological vector space, is necessarily continuous.

PROOF. We verify these assertions for real vector spaces, leaving the complex case to the exercises. The map $T : \mathbb{R}^n \to X$ in part 1 is obviously linear, 1-1 and onto. Also, it is continuous by part b of Exercise 3.2. Let us show that T^{-1} is continuous. Thus, let the net $\{x^\alpha\} = \{\sum_{i=1}^n t_i^\alpha x_i\}$ converge to 0 in X. Suppose, by way of contradiction, that there exists an i for which the net $\{t_i^\alpha\}$ does not converge to 0. Then let $\{t_i^{\alpha^\beta}\}$ be a subnet for which $\lim_\beta t_i^{\alpha^\beta} = t$, where t either is $\pm\infty$ or is a nonzero real number. Write $x^\alpha = t_i^\alpha x_i + x'^\alpha$. Then

$$(1/t_i^{\alpha^\beta})x^{\alpha^\beta} = x_i + (1/t_i^{\alpha^\beta})x'^{\alpha^\beta},$$

whence,

$$x_i = -\lim_\beta (1/t_i^{\alpha^\beta})x'^{\alpha^\beta},$$

implying that x_i belongs to the (closed) subspace spanned by the vectors

$$x_1, \ldots, x_{i-1}, x_{i+1}, \ldots, x_n,$$

and this is a contradiction, since the x_i's form a basis of X. Therefore, each of the nets $\{t_i^\alpha\}$ converges to 0, and T^{-1} is continuous.

We leave the proofs of parts 2 and 3 to the exercises.

EXERCISE 3.3. (a) Prove parts 2 and 3 of the preceding theorem in the case that X is a real topological vector space.

(b) Prove the preceding theorem in the case that X is a complex topological vector space.

EXERCISE 3.4. (Quotient Topological Vector Spaces) Let M be a linear subspace of a topological vector space X.

(a) Prove that the natural map π, which sends $x \in X$ to $x + M \in X/M$, is continuous and is an open map, where X/M is given the quotient topology.

(b) Show that X/M, equipped with the quotient topology, is a topological vector space if and only if M is a closed subspace of X. HINT: Use part e of Exercise 3.2.

(c) Suppose M is not closed in X. Show that, if U is any neighborhood of $0 \in X$, then $U + M$ contains the closure $\overline{M}$ of M.

(d) Conclude from part c that, if M is dense in X, then the only open subsets of X/M are X/M and $\emptyset$.

THEOREM 3.3. *Let X be a real topological vector space. Then X is locally compact if and only if X is finite dimensional.*

PROOF. If X is finite dimensional it is clearly locally compact, since the only topology on $\mathbb{R}^n$ is the usual Euclidean one. Conversely, suppose

U is a compact neighborhood of $0 \in X$, and let V be a neighborhood of 0 for which $V + V \subseteq U$. Because U is compact, there exists a finite set $x_1, \ldots, x_n$ of points in U such that

$$U \subseteq \cup_{i=1}^{n}(x_i + V).$$

Let M denote the subspace of X spanned by the points $x_1, \ldots, x_n$. Then M is a closed subspace, and the neighborhood $\pi(U)$ of 0 in X/M equals $\pi(V)$. Indeed, if $\pi(y) \in \pi(U)$, with $y \in U$, then there exists an $1 \leq i \leq n$ such that $y \in x_i + V$, whence $\pi(y) \in \pi(V)$.

It then follows that

$$\pi(U) = \pi(U) + \pi(U) = N\pi(U)$$

for every positive integer N, which implies that $\pi(U) = X/M$. So X/M is compact and hence is $\{0\}$. Therefore, $X = M$, and X is finite dimensional.

THEOREM 3.4. *Let T be a linear transformation of a real topological vector space X into a real topological vector space Y, and let M be the kernel of T. If π denotes the quotient map of X onto X/M, and if S is the unique linear transformation of the vector space X/M into Y satisfying $T = S \circ \pi$, then S is continuous if and only if T is continuous, and S is an open map if and only if T is an open map.*

PROOF. Since π is continuous and is an open map, see Exercise 3.4, It follows that T is continuous or open if S is continuous or open. If T is continuous, and if U is an open subset of Y, then $S^{-1}(U) = \pi(T^{-1}(U))$, and this is open because T is continuous and π is an open map. Hence, S is continuous.

Finally, if T is an open map and U is an open subset of X/M, then $S(U) = S(\pi(\pi^{-1}(U))) = T(\pi^{-1}(U))$, which is open because T is an open map and π is continuous. So, S is an open map.

THEOREM 3.5. (Characterization of Continuity) *If T is a linear transformation of a real (or complex) topological vector space X into $\mathbb{R}^n$ (or $\mathbb{C}^n$), then T is continuous if and only if $\ker(T)$ is closed. Further, T is continuous if and only if there exists a neighborhood of 0 in X on which T is bounded. If f is a linear functional on X, then f is continuous if and only if there exists a neighborhood of 0 on which f either is bounded above or is bounded below.*

PROOF. Suppose that X is a real vector space. If $M = \ker(T)$ is closed, and if $T = S \circ \pi$, then T is continuous because S is, X/M being finite dimensional. The converse is obvious.

If T is not continuous, then, from the preceding paragraph, M is not closed. So, by part c of Exercise 3.4, every neighborhood U of 0 is such that $U + M$ contains $\overline{M}$. If x is an element of $\overline{M} - M$, then $T(x) \neq 0$. Also, for any scalar λ, $\lambda x \in \overline{M} \subseteq U + M$, whence there exists an $m \in M$ such that $\lambda x - m \in U$. But then, $T(\lambda x - m) = \lambda T(x)$, showing that T is not bounded on U. Again, the converse is immediate.

The third claim of this theorem follows in the same manner as the second, and the complex cases for all parts are completely analogous to the real ones.

REMARK. We shall see that the graph of a linear transformation is important vis a vis the continuity of T. The following exercise demonstrates the initial aspects of this connection.

EXERCISE 3.5. (Continuity and the Graph) Let X and Y be topological vector spaces, and let T be a linear transformation from X into Y.

(a) Show that if T is continuous then the graph of T is a closed subspace of $X \times Y$.

(b) Let X and Y both be the normed linear space c_c (see part g of Exercise 3.2), and define T by $T(\{a_j\}) = \{ja_j\}$. Verify that the graph of T is a closed subset of $X \times Y$ but that T is not continuous.

(c) Show that, if the graph of T is closed, then the kernel of T is closed.

(d) Let $Y = \mathbb{R}^n$ or $\mathbb{C}^n$. Show that T is continuous if and only if the graph of T is closed.

EXERCISE 3.6. (a) Let T be a linear transformation from a normed linear space X into a normed linear space Y. Show that T is continuous if and only if there exists a constant M such that

$$\|T(x)\| \leq M\|x\|$$

for every $x \in X$.

(b) Let X be an infinite dimensional normed linear space. Prove that there exists a discontinuous linear functional on X. HINT: Show that there exists an infinite set of linearly independent vectors of norm 1. Then, define a linear functional that is not bounded on any neighborhood of 0.

(c) Show that, if $1 \leq p < \infty$, then $L^p(\mathbb{R})$ is a separable normed linear space. What about $L^\infty(\mathbb{R})$?

(d) Let μ be counting measure on an uncountable set X. Show that each $L^p(\mu)$ $(1 \leq p \leq \infty)$ is a normed linear space but that none is separable.

(e) Let Δ be a second-countable locally compact topological space. Show that $X = C_0(\Delta)$ is a separable normed linear space, where the norm on X is the supremum norm. (See Exercise 1.9.)

DEFINITION. Let X be a set, and let $\{f_\nu\}$ be a collection of real-valued (or complex-valued) functions on X. The *weak topology* on X, *generated* by the f_ν's, is the smallest topology on X for which each f_ν is continuous. A basis for this topology consists of sets of the form

$$V = \cap_{i=1}^n f_{\nu_i}^{-1}(U_i),$$

where each U_i is an open subset of $\mathbb{R}$ (or $\mathbb{C}$).

EXERCISE 3.7. (Vector Space Topology Generated by a Set of Linear Functionals) Let X be a real vector space and let $\{f_\nu\}$ be a collection of linear functionals on X that separates the points of X. Let $Y = \prod_\nu \mathbb{R}$, and define a function $F : X \to Y$ by $[F(x)](\nu) = f_\nu(x)$.

(a) Show that F is 1-1, and that with respect to the weak topology on X, generated by the f_ν's, F is a homeomorphism of X onto the subset $F(X)$ of Y. HINT: Compare the bases for the two topologies.

(b) Conclude that convergence in the weak topology on X, generated by the f_ν's, is described as follows:

$$x = \lim_\alpha x_\alpha \equiv f_\nu(x) = \lim_\alpha f_\nu(x_\alpha)$$

for all ν.

(c) Prove that X, equipped with the weak topology generated by the f_ν's, is a topological vector space.

(d) Show that Y is metrizable, and hence this weak topology on X is metrizable, if the set of f_ν's is countable.

(e) Verify that parts a through d hold if X is a complex vector space and each f_ν is a complex linear functional.

An important kind of topological vector space is obtained as a generalization of the preceding exercise, and is constructed as follows. Let X be a (real or complex) vector space, and let $\{\rho_\nu\}$ be a collection of semi-norms on X that separates the nonzero points of X from 0 in the sense that for each $x \neq 0$ there exists a ν such that $\rho_\nu(x) > 0$. For each $y \in X$ and each index ν, define $g_{y,\nu}(x) = \rho_\nu(x - y)$. Then X, equipped with the weakest topology making all of the $g_{y,\nu}$'s continuous, is a topological vector space, i.e., is Hausdorff and addition and scalar multiplication are continuous. A net $\{x_\alpha\}$ of elements in X converges in this topology to an element x if and only if $\rho_\nu(x - x_\alpha)$ converges to 0 for every ν. Further, this topology is a metrizable topology if the collection $\{\rho_\nu\}$ is countable.

We call this the *vector space topology* on X generated by the seminorms $\{\rho_\nu\}$ and denote this topological vector space by $(X, \{\rho_\nu\})$.

If $\rho_1, \rho_2, \ldots$ is a sequence of norms on X, then we call the topological vector space $(X, \{\rho_n\})$ a *countably normed* space.

EXERCISE 3.8. (Vector Space Topology Generated by a Set of Seminorms) Let X be a real (or complex) vector space and let $\{\rho_\nu\}$ be a collection of seminorms on X that separates the nonzero points of X from 0 in the sense that for each $x \neq 0$ there exists a ν such that $\rho_\nu(x) > 0$. For each $y \in X$ and each index ν, define $g_{y,\nu}(x) = \rho_\nu(x - y)$. Finally, let $\mathcal{T}$ be the topology on X generated by the $g_{y,\nu}$'s.

(a) Let x be an element of X and let V be an open set containing x. Show that there exist indices $\nu_1, \ldots, \nu_n$, elements $y_1, \ldots, y_n \in X$, and open sets $U_1, \ldots, U_n \subseteq \mathbb{R}$ ($\mathbb{C}$) such that

$$x \in \cap_{i=1}^n g_{y_i,\nu_i}^{-1}(U_i) \subseteq V.$$

(b) Conclude that convergence in the topology on X generated by the $g_{y,\nu}$'s is described by

$$x = \lim_\alpha x_\alpha \equiv \lim_\alpha \rho_\nu(x - x_\alpha) = 0$$

for each ν.

(c) Prove that X, equipped with the topology generated by the $g_{y,\nu}$'s, is a topological vector space. (HINT: Use nets.) Show further that this topology is metrizable if the collection $\{\rho_\nu\}$ is countable, i.e., if $\rho_1, \rho_2, \ldots$ is a sequence of seminorms. (HINT: Use the formula

$$d(x, y) = \sum_{n=1}^\infty 2^{-n} \min(\rho_n(x - y), 1).$$

Verify that d is a translation-invariant metric and that convergence with respect to this metric is equivalent to convergence in the topology $\mathcal{T}$.)

(d) Let X be a vector space, and let $\rho_1, \rho_2, \ldots$ be a sequence of seminorms that separate the nonzero points of X from 0. For each $n \geq 1$, define $p_n = \max_{k \leq n} \rho_k$. Prove that each p_n is a seminorm on X, that

$p_n \leq p_{n+1}$ for all n, and that the two topological vector spaces $(X, \{\rho_n\})$ and $(X, \{p_n\})$ are topologically isomorphic.

(e) Let X and $\{p_n\}$ be as in part d. Show that if V is a neighborhood of 0, then there exists an integer n and an $\epsilon > 0$ such that if $p_n(x) < \epsilon$, then $x \in V$. Deduce that, if f is a continuous linear functional on $(X, \{p_n\})$, then there exists an integer n and a constant M such that $|f(x)| \leq M p_n(x)$ for all $x \in X$.

(f) Let X be a normed linear space, and define $\rho(x) = \|x\|$. Prove that the topology on X determined by the norm coincides with the vector space topology generated by ρ.

EXERCISE 3.9. (a) Let X be the complex vector space of all infinitely differentiable complex-valued functions on $\mathbb{R}$. For each nonnegative integer n, define ρ_n on X by

$$\rho_n(f) = \sup_{|x| \leq n} \sup_{0 \leq i \leq n} |f^{(i)}(x)|,$$

where $f^{(i)}$ denotes the ith derivative of f. Show that the ρ_n's are seminorms (but not norms) that separate the nonzero points of X from 0, whence X is a metrizable complex topological vector space in the weak vector space topology generated by the ρ_n's. This vector space is usually denoted by $\mathcal{E}$.

(b) Let X be the complex vector space $C_0(\Delta)$, where Δ is a locally compact Hausdorff space. For each $\delta \in \Delta$, define ρ_δ on X by

$$\rho_\delta(f) = |f(\delta)|.$$

Show that, with respect to the weak vector space topology generated by the ρ_δ's, convergence is pointwise convergence of the functions.

EXERCISE 3.10. (Schwartz Space) Let $\mathcal{S}$ denote the set of all C^∞ complex-valued functions f on $\mathbb{R}$ that are rapidly decreasing, i.e., such that $x^n f^{(j)}(x) \in C_0(\mathbb{R})$ for every pair of nonnegative integers n and j. In other words, f and all its derivatives tend to 0 at $\pm\infty$ faster than the reciprocal of any polynomial.

(a) Show that every C^∞ function having compact support belongs to $\mathcal{S}$, and verify that $f(x) = x^k e^{-x^2}$ belongs to $\mathcal{S}$ for every integer $k \geq 0$.

(b) Show that $\mathcal{S}$ is a complex vector space, that each element of $\mathcal{S}$ belongs to every L^p space, and that $\mathcal{S}$ is closed under differentiation and multiplication by polynomials. What about antiderivatives of elements of $\mathcal{S}$? Are they again in $\mathcal{S}$?

(c) For each nonnegative integer n, define p_n on $\mathcal{S}$ by

$$p_n(f) = \sup_x \max_{0 \le i,j \le n} |x^j f^{(i)}(x)|.$$

Show that each p_n is a norm on $\mathcal{S}$, that $p_n(f) \le p_{n+1}(f)$ for all $f \in \mathcal{S}$, and that the topological vector space $(\mathcal{S}, \{p_n\})$ is a countably normed space. This countably normed vector space is called *Schwartz space*.

(d) Show that $f = \lim f_k$ in $\mathcal{S}$ if and only if $\{x^j f_k^{(i)}(x)\}$ converges uniformly to $x^j f^{(i)}(x)$ for every i and j.

(e) Prove that the map $f \to f'$ is a continuous linear transformation from $\mathcal{S}$ into itself. Is this transformation onto?

We introduce next a concept that is apparently purely from algebraic linear space theory and one that is of extreme importance in the topological aspect of Functional Analysis.

DEFINITION. A subset S of a vector space is called *convex* if $(1 - t)x + ty \in S$ whenever $x, y \in S$ and $0 \le t \le 1$. The *convex hull* of a set S is the smallest convex set containing S (the intersection of all convex sets containing S). A topological vector space X is called *locally convex* if there exists a neighborhood basis at 0 consisting of convex subsets of X. That is, if U is any neighborhood of 0 in X, then there exists a convex open set V such that $0 \in V \subseteq U$.

EXERCISE 3.11. (a) Let X be a real vector space. Show that the intersection of two convex subsets of X and the sum of two convex subsets of X is a convex set. If S is a subset of X, show that the intersection of all convex sets containing S is a convex set. Show also that the closure of a convex set is convex.

(b) Prove that a normed linear space is locally convex by showing that each ball centered at 0 in X is a convex set.

(c) Let X be a vector space and let $\{f_\nu\}$ be a collection of linear functionals on X that separates the points of X. Show that X, equipped with the weakest topology making all of the f_ν's continuous, is a locally convex topological vector space. (See Exercise 3.7.)

(d) Let X be a vector space, and let $\{\rho_\nu\}$ be a collection of seminorms on X that separates the nonzero points of X from 0. Prove that X, equipped with the weak vector space topology generated by the ρ_ν's, is a locally convex topological vector space. (See Exercise 3.8.)

(e) Suppose X is a locally convex topological vector space and that M is a subspace of X. Show that M is a locally convex topological vector space with respect to the relative topology. If M is a closed subspace of X, show that the quotient space X/M is a locally convex topological vector space.

(f) Show that all the L^p spaces are locally convex as well as the spaces $C_0(\Delta)$ under pointwise convergence, $\mathcal{E}$, and $\mathcal{S}$ of Exercises 3.9 and 3.10.

If X is a real vector space, recall that a function $\rho : X \to \mathbb{R}$ is called a *subadditive functional* if

(1) $\rho(x + y) \leq \rho(x) + \rho(y)$ for all $x, y \in X$.
(2) $\rho(tx) = t\rho(x)$ for all $x \in X$ and $t \geq 0$.

THEOREM 3.6. (Convex Neighborhoods of 0 and Continuous Subadditive Functionals) *Let X be a real topological vector space. If ρ is a continuous subadditive functional on X, then $\rho^{-1}(-\infty, 1)$ is a convex neighborhood of 0 in X. Conversely, if U is a convex neighborhood of 0, then there exists a continuous nonnegative subadditive functional ρ such that $\rho^{-1}(-\infty, 1) \subseteq U \subseteq \rho^{-1}(-\infty, 1]$. In addition, if U is symmetric, then ρ may be chosen to be a seminorm.*

PROOF. If ρ is a continuous subadditive functional, then it is immediate that $\rho^{-1}(-\infty, 1)$ is open, contains 0, and is convex.

Conversely, if U is a convex neighborhood of 0, define ρ on X by

$$\rho(x) = \frac{1}{\sup_{t>0, tx \in U} t} = \inf_{r>0, x \in rU} r.$$

(We interpret $\rho(x)$ as 0 if the supremum in the denominator is ∞, i.e., if $x \in rU$ for all $r > 0$.) Because U is an open neighborhood of $0 = 0 \times x$, and because scalar multiplication is continuous, the supremum in the above formula is always > 0, so that $0 \leq \rho(x) < \infty$ for every x. Notice also that if $t > 0$ and $t \times x \in U$, then $1/t \geq \rho(x)$.

It follows immediately that $\rho(rx) = r\rho(x)$ if $r \geq 0$, and, if U is symmetric, then $\rho(rx) = |r|\rho(x)$ for arbitrary real r.

If x and y are in X and $\epsilon > 0$ is given, choose real numbers t and s such that $tx \in U$, $sy \in U$, $1/t \leq \rho(x) + \epsilon$, and $1/s \leq \rho(y) + \epsilon$. Because U is convex, we have that

$$\frac{s}{t+s} tx + \frac{t}{t+s} sy = \frac{st}{t+s}(x+y) \in U.$$

Therefore, $\rho(x + y) \leq (s + t)/st$, whence

$$\rho(x + y) \leq (t + s)/st = (1/t) + (1/s) \leq \rho(x) + \rho(y) + 2\epsilon,$$

completing the proof that ρ is a subadditive functional in general and a seminorm if U is symmetric.

If $\rho(x) < 1$, then there exists a $t > 1$ so that $tx \in U$. Since U is convex, it then follows that $x \in U$. Also, if $x = 1 \times x \in U$, then $\rho(x) \leq 1$. Hence, $\rho^{-1}(-\infty, 1) \subseteq U \subseteq \rho^{-1}(-\infty, 1]$.

Finally, $\rho^{-1}(-\infty, \epsilon) \subseteq \epsilon U \subseteq \rho^{-1}(-\infty, \epsilon]$ for every positive ϵ, which shows that ρ is continuous at 0 and hence everywhere.

REMARK. The subadditive functional ρ constructed in the preceding proof is called the *Minkowski functional* associated to the convex neighborhood U.

THEOREM 3.7. (Hahn-Banach Theorem, Locally Convex Version) *Let X be a real locally convex topological vector space, let Y be a subspace of X, and let f be a continuous linear functional on Y with respect to the relative topology. Then there exists a continuous linear functional g on X whose restriction to Y is f.*

PROOF. By Theorem 3.5, there exists a neighborhood V of 0 in Y on which f is bounded, and by scaling we may assume that it is bounded by 1; i.e., $|f(y)| \leq 1$ if $y \in V$. Let W be a neighborhood of 0 in X such that $V = W \cap Y$, and let U be a symmetric convex neighborhood of 0 in X such that $U \subseteq W$. Let ρ be the continuous seminorm (Minkowski functional) on X associated to U as in the preceding theorem.

Now, if $y \in Y$, $t > 0$, and $ty \in U$, then

$$|f(y)| = (1/t)|f(ty)| \leq 1/t,$$

whence, by taking the supremum over all such t's,

$$|f(y)| \leq \rho(y),$$

showing that f is bounded by ρ on Y. Using Theorem 2.2, let g be a linear functional on X that extends f and such that $|g(x)| \leq \rho(x)$ for all $x \in X$. Then g is an extension of f and is continuous, so the proof is complete.

EXERCISE 3.12. Let M be a subspace of a locally convex topological vector space X. Prove that M is dense in X if and only if the only continuous linear functional f on X that is identically 0 on M is the 0 functional.

THEOREM 3.8. (Local Convexity and Existence of Continuous Linear Functionals) *A locally convex topological vector space has sufficiently many continuous linear functionals to separate its points.*

PROOF. Assume first that X is a real topological vector space. We will apply the Hahn-Banach Theorem. Suppose that $x \neq y$ are elements of X. Let U be a convex neighborhood of 0 which does not contain $y - x$, and let ρ be the Minkowski functional associated to U. Then $\rho(y-x) \geq 1$. Let Y be the subspace of X consisting of the real multiples of the nonzero vector $y - x$. Define a linear functional f on Y by

$$f(t(y - x)) = t.$$

Since $-\rho(-z) \leq \rho(z)$ for all $z \in Y$, observe that

$$f(z) \leq \rho(z)$$

for all $z \in Y$. By part c of Exercise 2.6, there exists a linear functional g on X which is an extension of f and for which $g(w) \leq \rho(w)$ for all $w \in X$. Since ρ is continuous at 0, and

$$-\rho(-x) \leq g(x) \leq \rho(x),$$

we see that g is continuous at 0 whence everywhere.

Now

$$g(y) - g(x) = g(y - x) = f(y - x) = 1 \neq 0,$$

showing that g separates the two points x and y.

Now, if X is a complex locally convex topological vector space, then it is obviously a real locally convex topological vector space. Hence, if $x \neq y$ are elements of X, then there exists a continuous real linear functional g on X such that $g(x) \neq g(y)$. But, as we have seen in Chapter I, the formula

$$f(z) = g(z) - ig(iz)$$

defines a complex linear functional on X, and clearly f is continuous and $f(x) \neq f(y)$.

EXERCISE 3.13. (Example of a Non-Locally-Convex Topological Vector Space) Let X be the vector space of all real-valued Lebesgue measurable functions on $[0, 1]$. For $f, g \in X$, set

$$d(f, g) = \sup_{\epsilon > 0} m(\{x : |f(x) - g(x)| \geq \epsilon\}),$$

where m denotes Lebesgue measure.

(a) Prove that d defines a translation-invariant metric on X. HINT: Show that $\{x : |f(x) - h(x)| \geq \epsilon\}$ is a subset of $\{x : |f(x) - g(x)| \geq \epsilon/2\} \cup \{x : |g(x) - h(x)| \geq \epsilon/2\}$.

(b) Show that convergence with respect to the metric d coincides with convergence in measure.

(c) Prove that, with respect to the topology determined by the metric d, X is a topological vector space, and that the subspace of measurable simple functions is dense in X.

(d) Let $\delta > 0$ be given. Show that if E is a measurable set of measure $< \delta$, then for every scalar c the function $c\chi_E$ belongs to the ball $B_\delta(0)$ of radius δ around $0 \in X$.

(e) Let f be a continuous linear functional on X, and let $B_\delta(0)$ be a neighborhood of 0 on which f is bounded. See Theorem 3.5. Show that $f(\chi_E) = 0$ for all E with $m(E) < \delta$, whence $f(\phi) = 0$ for every simple function $\phi \in X$.

(f) Conclude that the only continuous linear functional on X is the zero functional, whence the topology on X is not locally convex.

THEOREM 3.9. (Separation Theorem) *Let C be a closed convex subset of a locally convex real topological vector space X, and let x be an element of X that is not in C. Then there exists a continuous linear functional ϕ on X and a real number s such that $\phi(c) \leq s < \phi(x)$ for all $c \in C$.*

PROOF. Again, we apply the Hahn-Banach Theorem. Let U be a neighborhood of 0 such that $x + U$ does not intersect C. Let V be a convex symmetric neighborhood of 0 for which $V + V \subseteq U$, and write C' for the open convex set $V + C$. Then $x + V \cap C' = \emptyset$. If y is an element of C', write W for the convex neighborhood $C' - y$ of 0, and observe that $(x - y + V) \cap W = \emptyset$. Let ρ be the continuous subadditive functional associated to W as in Theorem 3.6. (ρ is not necessarily a seminorm since W need not be symmetric.) If Y is the linear span of the nonzero vector $x - y$, let f be defined on Y by $f(t(x - y)) = t\rho(x - y)$. Then f is a linear functional on Y satisfying $f(z) \leq \rho(z)$ for all $z \in Y$. By part c of Exercise 2.6, there exists a linear functional ϕ on X, which is an extension of f and which satisfies $\phi(w) \leq \rho(w)$ for all $w \in X$.

Since ρ is continuous, it follows that ϕ is continuous. Also, by the definition of ρ, if $z \in W$, then $\rho(z) \leq 1$, whence $\phi(z) \leq 1$. Now $\rho(x-y) > 1$. For, if t is sufficiently close to 1, then $t(x - y) \in x - y + V$, whence $t(x - y) \notin W$, and $\rho(x - y) \geq 1/t > 1$. So, $\phi(x - y) = f(x - y) =$

$\rho(x - y) > 1$. Setting $s = \phi(y) + 1$, we have $\phi(c) \leq s$ for all $c \in C$, and $\phi(x) > s$, as desired.

DEFINITION. Let C be a convex subset of a real vector space X. We say that a nonempty convex subset F of C is a *face* of C if: Whenever $x \in F$ is a *proper convex combination* of points in C (i.e., $x = (1 - t)y + tz$, with $y \in C$, $z \in C$, and $0 < t < 1$,) then both y and z belong to F.

A point $x \in C$ is called an *extreme point* of C if: Whenever $x = (1 - t)y + tz$, with $y \in C$, $z \in C$, and $0 < t < 1$, then $y = z = x$.

EXERCISE 3.14. (a) Let C be the closed unit ball in $L^p(\mathbb{R})$, for $1 < p < \infty$. Show that the extreme points of C are precisely the elements of the unit sphere, i.e., the elements f for which $\|f\|_p = 1$. HINT: Use the fact that $|(1 - t)y + tz|^p < (1 - t)|y|^p + t|z|^p$ if $y \neq z$ and $0 < t < 1$.

(b) If C is the closed unit ball in $L^1(\mathbb{R})$, show that C has no extreme points.

(c) Find the extreme points of the closed unit ball in $l^\infty(\mathbb{R})$.

(d) Find all the faces of a right circular cylinder, a tetrahedron, a sphere. Are all these faces closed sets?

(e) Suppose C is a closed convex set. Is the closure of a face of C again a face? Is every face of C necessarily closed?

(f) Show that a singleton, which is a face of a convex set C, is an extreme point of C. Show further that if a continuous linear functional attains a maximum (or minimum) on C, then it attains this extreme value at some extreme point.

(g) Show that the intersection of two faces of C is a face of C. Also, if ϕ is a linear functional on X, and $\max_{x \in C} \phi(x) = c$, show that $\phi^{-1}(c) \cap C$ is a face of C.

EXERCISE 3.15. (Hahn-Banach Theorem, Extreme Point Version) Let X be a real vector space, and let ρ be a seminorm (or subadditive functional) on X. If Z is a subspace of X, define F_Z to be the set of all linear functionals f on Z for which $f(z) \leq \rho(z)$ for all $z \in Z$.

(a) Prove that F_Z is a convex set of linear functionals.

(b) Let Y be a subspace of X. If f is an extreme point of F_Y, show that there is an extreme point $g \in F_X$ that is an extension of f. HINT: Mimic the proof of Theorem 2.2. That is, use the Hausdorff maximality principle to find a maximal pair (Z, h), for which h is an extension of f and h is an extreme point of F_Z. Then, following the notation in the proof to Theorem 2.2, show that $Z = X$ by choosing c to equal b.

We give two main theorems concerning the set of extreme points of a convex set.

THEOREM 3.10. (Krein-Milman Theorem) *Let C be a nonempty compact convex subset of a locally convex real topological vector space X. Then*

(1) *There exists an extreme point of C.*
(2) *C is the closure of the convex hull of its extreme points.*

PROOF. Let $\mathcal{F}$ be the collection of all closed faces of C, and consider $\mathcal{F}$ to be a partially ordered set by defining $F \leq F'$ if $F' \subseteq F$. Then, $\mathcal{F}$ is nonempty (C is an element of $\mathcal{F}$), and we let $\{F_\alpha\}$ be a maximal linearly ordered subset of $\mathcal{F}$ (the Hausdorff maximality principle). We set $F = \cap F_\alpha$, and note, since C is compact, that F is a nonempty closed (compact) face of C. We claim that F is a singleton, whence an extreme point of C. Indeed, if $x \in F$, $y \in F$, and $x \neq y$, let ϕ be a continuous linear functional which separates x and y, and let z be a point in the compact set F at which ϕ attains its maximum on F. Let $H = \phi^{-1}(\phi(z))$, and let $F' = F \cap H$. Then F' is a closed face of C which is properly contained in F. See the preceding exercise. But then the subset of $\mathcal{F}$, consisting of the F_α's together with F', is a strictly larger linearly ordered subset of $\mathcal{F}$, and this is a contradiction. Therefore, F is a singleton, and part 1 is proved.

Next, let C' be the closure of the convex hull of the extreme points of C. Then $C' \subseteq C$. If there is an $x \in C$ which is not in C', then, using the Separation Theorem (Theorem 3.9), let s be a real number and ϕ be a continuous linear functional for which $\phi(y) \leq s < \phi(x)$ for all $y \in C'$. Because C is compact and ϕ is continuous, there exists a $z \in C$ such that $\phi(z) \geq \phi(w)$ for all $w \in C$, and we let $C'' = C \cap \phi^{-1}(\phi(z))$. Then C'' is a nonempty compact convex subset of C, and $C' \cap C'' = \emptyset$. By part 1, there exists an extreme point p of C''. We claim that p is also an extreme point of C. Thus, if $p = (1 - t)q + tr$, with $q \in C$, $r \in C$, and $0 < t < 1$, then

$$\begin{aligned}
\phi(z) &= \phi(p) \\
&= (1 - t)\phi(q) + t\phi(r) \\
&\leq (1 - t)\phi(z) + t\phi(z) \\
&= \phi(z).
\end{aligned}$$

Therefore, $\phi(q) = \phi(r) = \phi(z)$, which implies that $q \in C''$ and $r \in C''$. Then, since p is an extreme point of C'', we have that $q = r = p$, as desired. But this implies that $p \in C'$, which is a contradiction. This completes the proof of part 2.

The Krein-Milman theorem is a topological statement about the set of extreme points of a compact convex set. Choquet's theorem, to follow, is a measure-theoretic statement about the set of extreme points of a compact convex set.

THEOREM 3.11. (Choquet Theorem) *Let X be a locally convex real topological vector space, let K be a metrizable, compact, convex subset of X, and let E denote the set of extreme points of K. Then:*

(1) *E is a Borel subset of K.*

(2) *For each $x \in K$, there exists a Borel probability measure μ_x on E such that*

$$f(x) = \int_E f(q) \, d\mu_x(q),$$

for every continuous linear functional f on X.

PROOF. Let A be the complement in $K \times K$ of the diagonal, i.e., the complement of the set of all pairs (x, x) for $x \in K$. Then A is an open subset of a compact metric space, and therefore A is a countable increasing union $A = \cup A_n$ of compact sets $\{A_n\}$. Define a function $I : (0, 1) \times A \to K$ by $I(t, y, z) = (1 - t)y + tz$. Then the range of I is precisely the complement of E in K. Also, since I is continuous, the range of I is the countable union of the compact sets $I([1/n, 1 - 1/n] \times A_n)$, whence the complement of E is an F_σ subset of K, so that E is a G_δ, hence a Borel set. This proves part 1.

Now, let Y denote the vector space of all continuous affine functions on K, i.e., all those continuous real-valued functions g on K for which

$$g((1 - t)y + tz) = (1 - t)g(y) + tg(z)$$

for all $y, z \in K$ and $0 \le t \le 1$. Note that the restriction to K of any continuous linear functional on X is an element of Y. Now Y is a subspace of $C(K)$. Since K is compact and metrizable, we have that $C(K)$ is a separable normed linear space in the uniform norm, whence Y is a separable normed linear space. Let $\{g_1, g_2, \ldots\}$ be a countable dense set in the unit ball $B_1(0)$ of Y, and define

$$g' = \sum_{i=1}^{\infty} 2^{-i} g_i^2.$$

Then g' is continuous on K, and is a proper convex function; i.e.,

$$g'((1 - t)y + tz) < (1 - t)g'(y) + tg'(z)$$

whenever $y, z \in K$, $y \neq z$, and $0 < t < 1$. Indeed, the series defining g' converges uniformly by the Weierstrass M test, showing that g' is continuous. Also, if $y, z \in K$, with $y \neq z$, there exists a continuous linear functional ϕ on X that separates y and z. In fact, any nonzero multiple of ϕ separates y and z. So, there exists at least one i such that $g_i(y) \neq g_i(z)$. Now, for any such i, if $0 < t < 1$, then

$$g_i^2((1-t)y + tz) < (1-t)g_i^2(y) + tg_i^2(z),$$

since

$$((1-t)a + tb)^2 - (1-t)a^2 - tb^2 < 0$$

for all $a \neq b$. Indeed, this function of b is 0 when $b = a$ and has a negative derivative for $b > a$. On the other hand, if i is such that $g_i(y) = g_i(z)$, then

$$g_i^2((1-t)y + tz) = (g_i((1-t)y + tz))^2 = g_i^2(y) = (1-t)g_i^2(y) + tg_i^2(z).$$

Hence,

$$g'((1-t)y + tz) = \sum_{i=1}^{\infty} 2^{-i}g_i^2((1-t)y + tz)$$

$$< \sum_{i=1}^{\infty} 2^{-i}[(1-t)g_i^2(y) + tg_i^2(z)]$$

$$= (1-t)g'(y) + tg'(z).$$

We let Y_1 be the linear span of Y and g', so that we may write each element of Y' as $g + rg'$, where $g \in Y$ and $r \in \mathbb{R}$.

Now, given an $x \in K$, define a function ρ_x on $C(K)$ by

$$\rho_x(h) = \inf c(x),$$

where the infimum is taken over all continuous concave functions c on K for which $h(y) \leq c(y)$ for all $y \in K$. Recall that a function c on K is called *concave* if

$$c((1-t)y + tz) \geq (1-t)c(y) + tc(z),$$

for all $y, z \in K$ and $0 \leq t \leq 1$. Because the sum of two concave functions is again concave and a positive multiple of a concave function is again concave, it follows directly that ρ_x is a subadditive functional on $C(K)$.

Note also that if c is a continuous concave function on K, then $\rho_x(c) = c(x)$. Define a linear functional ψ_x on Y_1 by

$$\psi_x(g + rg') = g(x) + r\rho_x(g').$$

Note that the identically 1 function I is an affine function, so it belongs to Y and hence to Y_1. It follows then that $\psi_x(I) = 1$. Also, we have that $\psi_x \leq \rho_x$ on Y_1 (see the exercise following), and we let ϕ_x be a linear functional on $C(K)$, which is an extension of ψ_x, and for which $\phi_x \leq \rho_x$ on $C(K)$. (We are using part c of Exercise 2.6.)

Note that, if $h \in C(K) \leq 0$, then $\rho_x(h) \leq 0$ (the 0 function is concave and $0 \geq h$), whence $\phi_x(h) \leq \rho_x(h) \leq 0$. It follows that ϕ_x is a positive linear functional. By the Riesz Representation Theorem, we let ν_x be the unique (finite) Borel measure on K for which

$$\phi_x(h) = \int h \, d\nu_x$$

for all $h \in C(K)$. Again letting I denote the identically 1 function on K, we have that

$$\begin{aligned}
\nu_x(K) &= \int I \, d\nu_x \\
&= \phi_x(I) \\
&= \psi_x(I) \\
&= 1,
\end{aligned}$$

showing that ν_x is a probability measure.

If f is a continuous linear functional on X, then

$$\int f \, d\nu_x = \phi_x(f) = \psi_x(f) = f(x),$$

since the restriction of f to K is a continuous affine function, whence in Y_1.

We prove next that ν_x is supported on E. To do this, let $\{c_n\}$ be a sequence of continuous concave functions on K for which $c_n \geq g'$ for all n and $\rho_x(g') = \lim c_n(x)$. Set $c = \liminf c_n$. Then c is a Borel function, hence is ν_x-measurable, and $c(y) \geq g'(y)$ for all $y \in K$. Hence,

$\int (c - g') \, d\nu_x \geq 0$. But,

$$\int (c - g') \, d\nu_x = \int (\liminf c_n - g') \, d\nu_x$$
$$\leq \liminf \int (c_n - g') \, d\nu_x$$
$$= \liminf \phi_x(c_n - g')$$
$$= \liminf \phi_x(c_n) - \phi_x(g')$$
$$= \liminf \phi_x(c_n) - \rho_x(g')$$
$$\leq \liminf \rho_x(c_n) - \rho_x(g')$$
$$= \liminf c_n(x) - \rho_x(g')$$
$$= \lim c_n(x) - \rho_x(g')$$
$$= 0.$$

Therefore, ν_x is supported on the set where c and g' agree. Let us show that $c(w) \neq g'(w)$ whenever $w \notin E$. Thus, if $w = (1 - t)y + tz$, for $y, z \in K$, $y \neq z$, and $0 < t < 1$, then

$$c(w) = \liminf c_n(w)$$
$$= \liminf c_n((1 - t)y + tz)$$
$$\geq \liminf[(1 - t)c_n(y) + tc_n(z)]$$
$$\geq (1 - t)g'(y) + tg'(z)$$
$$> g'((1 - t)y + tz)$$
$$= g'(w).$$

Define μ_x to be the restriction of ν_x to E. Then μ_x is a Borel probability measure on E, and

$$\int_E f \, d\mu_x = \int_K f \, d\nu_x = f(x)$$

for all continuous linear functionals f on X. This completes the proof.

EXERCISE 3.16. (a) Verify that the function ρ_x in the preceding proof is a subadditive functional and that $\psi_x(h) \leq \rho_x(h)$ for all $h \in Y_1$.

(b) Let $X = \mathbb{R}^2$, let $K = \{(s, t) : |s| + |t| \leq 1\}$, and let $x = (0, 0)$ be the origin. Show that there are uncountably many different Borel probability measures μ on the set E of extreme points of K for which $f(x) = \int_E f(q) \, d\mu(q)$ for all linear functionals on X. Conclude that there can be no uniqueness assertion in Choquet's Theorem.

CHAPTER IV

NORMED LINEAR SPACES AND BANACH SPACES

DEFINITION A *Banach space* is a real normed linear space that is a complete metric space in the metric defined by its norm. A *complex Banach space* is a complex normed linear space that is, as a real normed linear space, a Banach space. If X is a normed linear space, x is an element of X, and δ is a positive number, then $B_\delta(x)$ is called the *ball of radius δ around x*, and is defined by $B_\delta(x) = \{y \in X : \|y - x\| < \delta\}$. The *closed ball* $\overline{B}_\delta(x)$ of radius δ around x is defined by $\overline{B}_\delta(x) = \{y \in X : \|y - x\| \leq \delta\}$. By B_δ and $\overline{B}_\delta$ we shall mean the (open and closed) balls of radius δ around 0.

Two normed linear spaces X and Y are *isometrically isomorphic* if there exists a linear isomorphism $T : X \to Y$ which is an isometry of X onto Y. In this case, T is called an *isometric isomorphism*.

If $X_1, \ldots X_n$ are n normed linear spaces, we define a norm on the (algebraic) direct sum $X = \bigoplus_{i=1}^{n} X_i$ by

$$\|(x_1, \ldots, x_n)\| = \max_{i=1}^{n} \|x_i\|.$$

This is frequently called the *max norm*.

Our first order of business is to characterize those locally convex topological vector spaces whose topologies are determined by a norm, i.e., those locally convex topological vector spaces that are normable.

DEFINITION. Let X be a topological vector space. A subset $S \subseteq X$ is called *bounded* if for each neighborhood W of 0 there exists a positive scalar c such that $S \subseteq cW$.

THEOREM 4.1. (Characterization of Normable Spaces) *Let X be a locally convex topological vector space. Then X is a normable vector space if and only if there exists a bounded convex neighborhood of 0.*

PROOF. If X is a normable topological vector space, let $\| \cdot \|$ be a norm on X that determines the topology. Then B_1 is clearly a bounded convex neighborhood of 0.

Conversely, let U be a bounded convex neighborhood of 0 in X. We may assume that U is symmetric, since, in any event, $U \cap (-U)$ is also bounded and convex. Let ρ be the seminorm (Minkowski functional) on X associated to U as in Theorem 3.6. We show first that ρ is actually a norm.

Thus, let $x \neq 0$ be given, and choose a convex neighborhood V of 0 such that $x \notin V$. Note that, if $tx \in V$, then $|t| < 1$. Choose $c > 0$ so that $U \subseteq cV$, and note that if $tx \in U$, then $tx \in cV$, whence $|t| < c$. Therefore, recalling the definition of $\rho(x)$,

$$\rho(x) = \frac{1}{\sup_{t>0, tx \in U} t},$$

we see that $\rho(x) \geq c > 0$, showing that ρ is a norm.

We must show finally that the given topology agrees with the one defined by the norm ρ. Since, by Theorem 3.6, ρ is continuous, it follows immediately that $B_\epsilon = \rho^{-1}(-\infty, \epsilon)$ is open in the given topology, showing that the topology defined by the norm is contained in the given topology. Conversely, if V is an open subset of the given topology and $x \in V$, let W be a neighborhood of 0 such that $x + W \subseteq V$. Choose $c > 0$ so that $U \subseteq cW$. Again using Theorem 3.6, we see that $B_1 = \rho^{-1}(-\infty, 1) \subseteq U \subseteq cW$, whence $B_{1/c} = \rho^{-1}(-\infty, (1/c)) \subseteq W$, and $x + B_{1/c} \subseteq V$. This shows that V is open in the topology defined by the norm. Q.E.D.

EXERCISE 4.1. (a) (Characterization of Banach Spaces) Let X be a normed linear space. Show that X is a Banach space if and only if every absolutely summable infinite series in X is summable in X. (An infinite series $\sum x_n$ is *absolutely summable* in X if $\sum \|x_n\| < \infty$.) HINT: If $\{y_n\}$ is a Cauchy sequence in X, choose a subsequence $\{y_{n_k}\}$ for which $\|y_{n_k} - y_{n_{k+1}}\| < 2^{-k}$.

(b) Use part a to verify that all the spaces $L^p(\mathbb{R})$, $1 \leq p \leq \infty$, are Banach spaces, as is $C_0(\Delta)$.

(c) If c_0 is the set of all sequences $\{a_n\}$, $n = 0, 1, \ldots$, satisfying $\lim a_n = 0$, and if we define $\|\{a_n\}\| = \max |a_n|$, show that c_0 is a Banach space.

(d) Let X be the set of all continuous functions on $[0,1]$, which are differentiable on $(0,1)$. Set $\|f\| = \sup_{x \in [0,1]} |f(x)|$. Show that X is a normed linear space but is not a Banach space.

(e) If $X_1, \ldots, X_n$ are normed linear spaces, show that the direct sum $\bigoplus_{i=1}^{n} X_i$, equipped with the max norm, is a normed linear space. If each X_i is a Banach space, show that $\bigoplus_{i=1}^{n} X_i$ is a Banach space.

(f) Let $X_1, \ldots, X_n$ be normed linear spaces. Let $x = (x_1, \ldots, x_n)$ be in $\bigoplus_{i=1}^{n} X_i$, and define $\|x\|_1$ and $\|x\|_2$ by

$$\|x\|_1 = \sum_{i=1}^{n} \|x_i\|,$$

and

$$\|x\|_2 = \sqrt{\sum_{i=1}^{n} \|x_i\|^2}.$$

Prove that both $\| \cdot \|_1$ and $\| \cdot \|_2$ are norms on $\bigoplus_{i=1}^{n} X_i$. Show further that

$$\|x\| \le \|x\|_2 \le \|x\|_1 \le n\|x\|.$$

(g) Let $\{X_i\}$ be an infinite sequence of nontrivial normed linear spaces. Prove that the direct product $\prod X_i$ is a metrizable, locally convex, topological vector space, but that there is no definition of a norm on $\prod X_i$ that defines its topology. HINT: In a normed linear space, given any bounded set A and any neighborhood U of 0, there exists a number t such that $A \subseteq tU$.

EXERCISE 4.2. (Schwartz Space $\mathcal{S}$ is Not Normable) Let $\mathcal{S}$ denote Schwartz space, and let $\{\rho_n\}$ be the seminorms (norms) that define the topology on $\mathcal{S}$:

$$\rho_n(f) = \sup_{x} \max_{0 \le i,j \le n} |x^j f^{(i)}(x)|.$$

(a) If V is a neighborhood of 0 in $\mathcal{S}$, show that there exists an integer n and an $\epsilon > 0$ such that $\rho_n^{-1}(-\infty, \epsilon) \subseteq V$; i.e., if $\rho_n(h) < \epsilon$, then $h \in V$.

(b) Given the nonnegative integer n from part a, show that there exists a C^∞ function g such that $g(x) = 1/x^{n+1/2}$ for $|x| \ge 2$. Note that

$$\sup_{x} \max_{0 \le i,j \le n} |x^j g^{(i)}(x)| < \infty.$$

(Of course, g is not an element of $\mathcal{S}$.)

(c) Let n be the integer from part a and let f be a C'^∞ function with compact support such that $|f(x)| \leq 1$ for all x and $f(0) = 1$. For each integer $M > 0$, define $g_M(x) = g(x)f(x - M)$, where g is the function from part b. Show that each $g_m \in \mathcal{S}$ and that there exists a positive constant c such that $\rho_n(g_M) < c$ for all M; i.e., $(\epsilon/c)g_m \in V$ for all m. Further, show that for each $M \geq 2$, $\rho_{n+1}(g_M) \geq \sqrt{M}$.

(d) Show that the neighborhood V of 0 from part a is not bounded in $\mathcal{S}$. HINT: Define W to be the neighborhood $\rho_{n+1}^{-1}(-\infty, 1)$, and show that no multiple of W contains V.

(e) Conclude that $\mathcal{S}$ is not normable.

THEOREM 4.2. (Subspaces and Quotient Spaces) Let X be a Banach space and let M be a closed linear subspace.

(1) M is a Banach space with respect to the restriction to M of the norm on X.

(2) If $x + M$ is a coset of M, and if $\|x + M\|$ is defined by

$$\|x + M\| = \inf_{y \in x+M} \|y\| = \inf_{m \in M} \|x + m\|,$$

then the quotient space X/M is a Banach space with respect to this definition of norm.

(3) The quotient topology on X/M agrees with the topology determined by the norm on X/M defined in part 2.

PROOF. M is certainly a normed linear space with respect to the restricted norm. Since it is a closed subspace of the complete metric space X, it is itself a complete metric space, and this proves part 1.

We leave it to the exercise that follows to show that the given definition of $\|x + M\|$ does make X/M a normed linear space. Let us show that this metric space is complete. Thus let $\{x_n + M\}$ be a Cauchy sequence in X/M. It will suffice to show that some subsequence has a limit in X/M. We may replace this Cauchy sequence by a subsequence for which

$$\|(x_{n+1} + M) - (x_n + M)\| = \|(x_{n+1} - x_n) + M\| < 2^{-(n+1)}.$$

Then, we may choose elements $\{y_n\}$ of X such that for each $n \geq 1$ we have

$$y_n \in (x_{n+1} - x_n) + M,$$

and $\|y_n\| < 2^{-(n+1)}$. We choose y_0 to be any element of $x_1 + M$. If $z_N = \sum_{n=0}^{N} y_n$, then it follows routinely that $\{z_N\}$ is a Cauchy sequence in X, whence has a limit z. We claim that $z + M$ is the limit of the sequence $\{x_N + M\}$. Indeed,

$$\|(z + M) - (x_N + M)\| = \|(z - x_N) + M\|$$
$$= \inf_{y \in (z - x_N) + M} \|y\|.$$

Since $z = \sum_{n=0}^{\infty} y_n$, and since $\sum_{n=0}^{N-1} y_n \in x_N + M$, It follows that $\sum_{n=N}^{\infty} y_n \in (z - x_N) + M$. Therefore,

$$\|(z + M) - (x_N + M)\| \leq \|\sum_{n=N}^{\infty} y_n\|$$
$$\leq \sum_{n=N}^{\infty} 2^{-(n+1)}$$
$$= 2^{-N},$$

completing the proof of part 2.

We leave part 3 to the exercise that follows.

EXERCISE 4.3. Let X and M be as in the preceding theorem.

(a) Verify that the definition of $\|x + M\|$, given in the preceding theorem, makes X/M into a normed linear space.

(b) Prove that the quotient topology on X/M agrees with the topology determined by the norm on X/M.

(c) Suppose X is a vector space, ρ is a seminorm on X, and $M = \{x : \rho(x) = 0\}$. Prove that M is a subspace of X. Define p on X/M by

$$p(x + M) = \inf_{m \in M} \rho(x + m).$$

Show that p is a norm on the quotient space X/M.

EXERCISE 4.4. (a) Suppose X and Y are topologically isomorphic normed linear spaces, and let S denote a linear isomorphism of X onto Y that is a homeomorphism. Prove that there exist positive constants C_1 and C_2 such that

$$\|x\| \leq C_1 \|S(x)\|$$

and

$$\|S(x)\| \leq C_2 \|x\|$$

for all $x \in X$. Deduce that, if two norms $\| \cdot \|_1$ and $\| \cdot \|_2$ determine identical topologies on a vector space X, then there exist constants C_1 and C_2 such that

$$\|x\|_1 \leq C_1 \|x\|_2 \leq C_2 \|x\|_1$$

for all $x \in X$.

(b) Suppose S is a linear transformation of a normed linear space X into a topological vector space Y. Assume that $S(\overline{B}_1)$ contains a neighborhood U of 0 in Y. Prove that S is an open map of X onto Y.

We come next to one of the important applications of the Baire category theorem in functional analysis.

THEOREM 4.3. (Isomorphism Theorem) *Suppose S is a continuous linear isomorphism of a Banach space X onto a Banach space Y. Then S^{-1} is continuous, and X and Y are topologically isomorphic.*

PROOF. For each positive integer n, let A_n be the closure in Y of $S(\overline{B}_n)$. Then, since S is onto, $Y = \cup A_n$. Because Y is a complete metric space, it follows from the Baire category theorem that some A_n, say A_N, must have nonempty interior. Therefore, let $y_0 \in Y$ and $\epsilon > 0$ be such that $B_\epsilon(y_0) \subset A_N$. Let $x_0 \in X$ be the unique element such that $S(x_0) = y_0$, and let k be an integer larger than $\|x_0\|$. Then A_{N+k} contains $A_N - y_0$, so that the closed set A_{N+k} contains $\overline{B}_\epsilon(0)$. This implies that if $w \in Y$ satisfies $\|w\| \leq \epsilon$, and if δ is any positive number, then there exists an $x \in X$ for which $\|S(x) - w\| < \delta$ and $\|x\| \leq N + k$. Write $M = (N + k)/\epsilon$. It follows then by scaling that, given any $w \in Y$ and any $\delta > 0$, there exists an $x \in X$ such that $\|S(x) - w\| < \delta$ and $\|x\| \leq M \|w\|$. We will use the existence of such an x recursively below.

We now complete the proof by showing that

$$\|S^{-1}(w)\| \leq 2M \|w\|$$

for all $w \in Y$, which will imply that S^{-1} is continuous. Thus, let $w \in Y$ be given. We construct sequences $\{x_n\}$, $\{w_n\}$ and $\{\delta_n\}$ as follows: Set $w_1 = w$, $\delta_1 = 1/2$, and choose x_1 so that $\|w_1 - S(x_1)\| < \delta_1$ and $\|x_1\| \leq M \|w_1\|$. Next, set $w_2 = w_1 - S(x_1)$, $\delta_2 = 1/4$, and choose x_2 such that $\|w_2 - S(x_2)\| < \delta_2$ and $\|x_2\| \leq M \|w_2\| \leq (M/2) \|w\|$. Continuing inductively, we construct the sequences $\{w_n\}, \{\delta_n\}$ and $\{x_n\}$ so that

$$w_n = w_{n-1} - S(x_{n-1}),$$

$$\delta_n = 1/2^n,$$

and x_n so that

$$\|w_n - S(x_n)\| < \delta_n$$

and

$$\|x_n\| \leq M\|w_n\| < (M/2^{n-1})\|w\|.$$

It follows that the infinite series $\sum x_n$ converges in X, its sequence of partial sums being a Cauchy sequence, to an element x and that $\|x\| \leq 2M\|w\|$. Also, $w_n = w - \sum_{i=1}^{n-1} S(x_i)$. So, since S is continuous and $0 = \lim w_n$, we have that $S(x) = S(\sum_{n=1}^{\infty} x_n) = \sum_{n=1}^{\infty} S(x_n) = w$. Finally,

$$\|S^{-1}(w)\| = \|x\| \leq 2M\|w\|,$$

and the proof is complete.

THEOREM 4.4. (Open Mapping Theorem) *Let T be a continuous linear transformation of a Banach space X onto a Banach space Y. Then T is an open map.*

PROOF. Since T is continuous, its kernel M is a closed linear subspace of X. Let S be the unique linear transformation of X/M onto Y satisfying $T = S \circ \pi$, where π denotes the natural map of X onto X/M. Then, by Theorems 3.4 and 4.2, S is a continuous isomorphism of the Banach space X/M onto the Banach space Y. Hence, S is an open map, whence T is an open map.

THEOREM 4.5. (Closed Graph Theorem) *Suppose T is a linear transformation of a Banach space X into a Banach space Y, and assume that the graph G of T is a closed subset of the product Banach space $X \times Y = X \oplus Y$. Then T is continuous.*

PROOF. Since the graph G is a closed linear subspace of the Banach space $X \oplus Y$, it is itself a Banach space in the restricted norm (max norm) from $X \oplus Y$. The map S from G to X, defined by $S(x, T(x)) = x$, is therefore a norm-decreasing isomorphism of G onto X. Hence S^{-1} is continuous by the Isomorphism Theorem. The linear transformation P of $X \oplus Y$ into Y, defined by $P(x, y) = y$, is norm-decreasing whence continuous. Finally, $T = P \circ S^{-1}$, and so is continuous.

EXERCISE 4.5. (a) Let X be the vector space of all continuous functions on $[0, 1]$ that have uniformly continuous derivatives on $(0, 1)$. Define a norm on X by $\|f\| = \sup_{0 < x < 1} |f(x)| + \sup_{0 < x < 1} |f'(x)|$. Let Y be the vector space of all uniformly continuous functions on $(0, 1)$, equipped with the norm $\|f\| = \sup_{0 < x < 1} |f(x)|$. Define $T : X \to Y$ by $T(f) = f'$.

Prove that X and Y are Banach spaces and that T is a continuous linear transformation.

(b) Now let X be the vector space of all continuous functions f on $[0, 1]$, for which $f(0) = 0$ and whose derivative f' is in L^p (for some fixed $1 \leq p \leq \infty$). Define a norm on X by $\|f\| = \|f\|_p$. Let $Y = L^p$, and define $T : X \to Y$ by $T(f) = f'$. Prove that T is not continuous, but that the graph of T is closed in $X \times Y$. How does this example relate to the preceding theorem?

(c) Prove analogous results to Theorems 4.3, 4.4, and 4.5 for Frechet spaces.

DEFINITION. Let X and Y be normed linear spaces. By $L(X, Y)$ we shall mean the set of all continuous linear transformations from X into Y. We refer to elements of $L(X, Y)$ as *operators* from X to Y. If $T \in L(X, Y)$, we define the *norm* of T, denoted by $\|T\|$, by

$$\|T\| = \sup_{\|x\| \leq 1} \|T(x)\|.$$

EXERCISE 4.6. Let X and Y be normed linear spaces.

(a) Let T be a linear transformation of X into Y. Verify that $T \in L(X, Y)$ if and only if

$$\|T\| = \sup_{\|x\| \leq 1} \|T(x)\| < \infty.$$

(b) Let T be in $L(X, Y)$. Show that the norm of T is the infimum of all numbers M for which $\|T(x)\| \leq M\|x\|$ for all $x \in X$.

(c) For each $x \in X$ and $T \in L(X, Y)$, show that $\|T(x)\| \leq \|T\|\|x\|$.

THEOREM 4.6. *Let X and Y be normed linear spaces.*

(1) *The set $L(X, Y)$ is a vector space with respect to pointwise addition and scalar multiplication. If X and Y are complex normed linear spaces, then $L(X, Y)$ is a complex vector space.*

(2) *$L(X, Y)$, equipped with the norm defined above, is a normed linear space.*

(3) *If Y is a Banach space, then $L(X, Y)$ is a Banach space.*

PROOF. We prove part 3 and leave parts 1 and 2 to the exercises. Thus, suppose Y is a Banach space, and let $\{T_n\}$ be a Cauchy sequence in $L(X, Y)$. Then the sequence $\{\|T_n\|\}$ is bounded, and we let M be a number for which $\|T_n\| \leq M$ for all n. For each $x \in X$, we have that

$\|(T_n(x) - T_m(x))\| \leq \|T_n - T_m\| \|x\|$, whence the sequence $\{T_n(x)\}$ is a Cauchy sequence in the complete metric space Y. Hence there exists an element $T(x) \in Y$ such that $T(x) = \lim T_n(x)$. This mapping T, being the pointwise limit of linear transformations, is a linear transformation, and it is continuous, since $\|T(x)\| = \lim \|T_n(x)\| \leq M\|x\|$. Consequently, T is an element of $L(X, Y)$.

We must show finally that T is the limit in $L(X, Y)$ of the sequence $\{T_n\}$. To do this, let $\epsilon > 0$ be given, and choose an N such that $\|T_n - T_m\| < \epsilon/2$ if $n, m \geq N$. If $x \in X$ and $\|x\| \leq 1$, then

$$\|T(x) - T_n(x)\| \leq \limsup_{m} \|T(x) - T_m(x)\| + \limsup_{m} \|T_m(x) - T_n(x)\|$$
$$\leq 0 + \limsup_{m} \|T_m - T_n\| \|x\|$$
$$\leq \epsilon/2,$$

whenever $n \geq N$. Since this is true for an arbitrary x for which $\|x\| \leq 1$, it follows that

$$\|T - T_n\| \leq \epsilon/2 < \epsilon$$

whenever $n \geq N$, as desired.

EXERCISE 4.7. Prove parts 1 and 2 of Theorem 4.6.

The next theorem gives another application to functional analysis of the Baire category theorem.

THEOREM 4.7. (Uniform Boundedness Principle) *Let X be a Banach space, let Y be a normed linear space, and suppose $\{T_n\}$ is a sequence of elements in $L(X, Y)$. Assume that, for each $x \in X$, the sequence $\{T_n(x)\}$ is bounded in Y. (That is, the sequence $\{T_n\}$ is pointwise bounded.) Then there exists a positive constant M such that $\|T_n\| \leq M$ for all n. (That is, the sequence $\{T_n\}$ is uniformly bounded.)*

PROOF. For each positive integer j, let A_j be the set of all $x \in X$ such that $\|T_n(x)\| \leq j$ for all n. Then each A_j is closed ($A_j = \cap_n T_n^{-1}(\overline{B_j})$), and $X = \cup A_j$. By the Baire category theorem, some A_j, say A_J, has nonempty interior. Let $\epsilon > 0$ and $x_0 \in X$ be such that A_J contains $B_\epsilon(x_0)$. If k is an integer for which $\|x_0\| < k$, then we see that A_{J+k} contains B_ϵ. Hence, if $\|z\| < \epsilon$, then $\|T_n(z)\| \leq J + k$ for all n.

Now, given a nonzero $x \in X$, we write $z = (\epsilon/2\|x\|)x$. So, for any n,

$$\|T_n(x)\| = (2\|x\|/\epsilon)\|T_n(z)\|$$
$$\leq (2\|x\|/\epsilon)(J + k)$$
$$= M\|x\|,$$

where $M = 2(J + k)/\epsilon$. It follows then that $\|T_n\| \leq M$ for all n, as desired.

THEOREM 4.8. *Let X be a Banach space, let Y be a normed linear space, let $\{T_n\}$ be a sequence of elements of $L(X, Y)$, and suppose that $\{T_n\}$ converges pointwise to a function $T : X \to Y$. Then T is a continuous linear transformation of X into Y; i.e., the pointwise limit of a sequence of continuous linear transformations from a Banach space into a normed linear space is continuous and linear.*

PROOF. It is immediate that the pointwise limit (when it exists) of a sequence of linear transformations is again linear. Since any convergent sequence in Y, e.g., $\{T_n(x)\}$, is bounded, it follows from the preceding theorem that there exists an M so that $\|T_n\| \leq M$ for all n, whence $\|T_n(x)\| \leq M\|x\|$ for all n and all $x \in X$. Therefore, $\|T(x)\| \leq M\|x\|$ for all x, and this implies that T is continuous.

EXERCISE 4.8. (a) Extend the Uniform Boundedness Principle from a sequence to a set S of elements of $L(X, Y)$.

(b) Restate the Uniform Boundedness Principle for a sequence $\{f_n\}$ of continuous linear functionals, i.e., for a sequence in $L(X, \mathbb{R})$ or $L(X, \mathbb{C})$.

(c) Let c_c denote the vector space of all sequences $\{a_j\}$, $j = 1, 2, \ldots$ that are eventually 0, and define a norm on c_c by

$$\|\{a_j\}\| = \max |a_j|.$$

Define a linear functional f_n on c_c by $f_n(\{a_j\}) = na_n$. Prove that the sequence $\{f_n\}$ is a sequence of continuous linear functionals that is pointwise bounded but not bounded in norm. Why doesn't this contradict the Uniform Boundedness Principle?

(d) Let c_c be as in part c. Define a sequence $\{f_n\}$ of linear functionals on c_c by $f_n(\{a_j\}) = \sum_{j=1}^{n} a_j$. Show that $\{f_n\}$ is a sequence of continuous linear functionals that converges pointwise to a discontinuous linear functional. Why doesn't this contradict Theorem 4.8?

(e) Let c_0 denote the Banach space of sequences $a_0, a_1, \ldots$ for which $\lim a_n = 0$, where the norm on c_0 is given by

$$\|\{a_n\}\| = \max |a_n|.$$

If $\alpha = \{n_1 < n_2 < \ldots < n_k\}$ is a finite set of positive integers, define f_α on c_0 by

$$f_\alpha(\{a_j\}) = f_{n_1,\ldots,n_k}(\{a_j\}) = n_1 a_{n_k}.$$

Show that each f_α is a continuous linear functional on c_0.

(f) Let D denote the set consisting of all the finite sets $\alpha = \{n_1 < n_2 < \ldots < n_k\}$ of positive integers. Using inclusion as the partial ordering on D, show that D is a directed set, and let $\{f_\alpha\}$ be the corresponding net of linear functionals, as defined in part e, on c_0. Show that $\lim_\alpha f_\alpha = 0$. Show also that the net $\{f_\alpha\}$ is not uniformly bounded in norm. Explain why this does not contradict part a of this exercise.

DEFINITION. A *Banach algebra* is a Banach space A on which there is also defined a binary operation of multiplication that is associative, distributive over addition, and satisfies $\|xy\| \leq \|x\|\|y\|$ for all $x, y \in A$.

EXERCISE 4.9. Let X be a Banach space. Using composition of transformations as a multiplication, show that $L(X, X)$ is a Banach algebra.

EXERCISE 4.10. Let X be the Banach space $\mathbb{R}^2$ with respect to the usual norm
$$\|x\| = \|(x_1, x_2)\| = \sqrt{x_1^2 + x_2^2},$$
and let $(1, 0)$ and $(0, 1)$ be the standard basis for X. Let T be an element of $L(X, X)$, and represent T by a 2×2 matrix $\begin{pmatrix} a & b \\ c & d \end{pmatrix}$. Compute the norm of T in terms of a, b, c, d. Can you do the same for $X = \mathbb{R}^3$?

EXERCISE 4.11. Let X be a normed linear space, let Y be a dense subspace of X, and let Z be a Banach space.

(a) If $T \in L(Y, Z)$, show that there exists a unique element $T' \in L(X, Z)$ such that the restriction of T' to Y is T. That is, T has a unique continuous extension to all of X.

(b) Show that the map $T \to T'$, of part a, is an isometric isomorphism of $L(Y, Z)$ onto $L(X, Z)$.

(c) Suppose $\{T_n\}$ is a uniformly bounded sequence of elements of $L(X, Z)$. Suppose that the sequence $\{T_n(y)\}$ converges for every $y \in Y$. Show that the sequence $\{T_n(x)\}$ converges for every $x \in X$.

EXERCISE 4.12. Let X be a normed linear space, and let $\overline{X}$ denote the completion of the metric space X (e.g., the space of equivalence classes of Cauchy sequences in X). Show that $\overline{X}$ is in a natural way a Banach space with X isometrically imbedded as a dense subspace.

THEOREM 4.9. (Hahn-Banach Theorem, Normed Linear Space Version) *Let Y be a subspace of a normed linear space X. Suppose f is a continuous linear functional on Y; i.e., $f \in L(Y, \mathbb{R})$. Then there exists* *a continuous linear functional g on X, i.e., an element of* $L(X, \mathbb{R})$, *such that*

(1) *g is an extension of f.*

(2) $\|g\| = \|f\|$.

PROOF. If ρ is defined on X by $\rho(x) = \|f\|\|x\|$, then ρ is a seminorm on X. Clearly,

$$f(y) \leq |f(y)| \leq \|f\|\|y\| = \rho(y)$$

for all $y \in Y$. By the seminorm version of the Hahn-Banach Theorem, there exists a linear functional g on X, which is an extension of f, such that $g(x) \leq \rho(x) = \|f\|\|x\|$, for all $x \in X$, and this implies that g is continuous, and $\|g\| \leq \|f\|$. Obviously $\|g\| \geq \|f\|$ since g is an extension of f.

EXERCISE 4.13. (a) Let X be a normed linear space and let $x \in X$. Show that $\|x\| = \sup_f f(x)$, where the supremum is taken over all continuous linear functionals f for which $\|f\| \leq 1$. Show, in fact, that this supremum is actually attained.

(b) Let $1 \leq p < \infty$, and let X be the complex Banach space $L^p(\mathbb{R})$. Let p' be such that $1/p + 1/p' = 1$, and let D be a dense subspace of $L^{p'}(\mathbb{R})$. If $f \in X$, show that

$$\|f\|_p = \sup_{\|g\|_{p'}=1} |\int f(x)g(x)\, dx|.$$

EXERCISE 4.14. Let X and Y be normed linear spaces, and let $T \in L(X, Y)$. Prove that the norm of T is given by

$$\|T\| = \sup_x \sup_f |f(T(x))|,$$

where the supremum is taken over all $x \in X$, $\|x\| \leq 1$ and all $f \in L(Y, \mathbb{R})$ for which $\|f\| \leq 1$.

We close this chapter with a theorem from classical analysis.

THEOREM 4.10. (Riesz Interpolation Theorem) *Let D be the linear space of all complex-valued measurable simple functions on* $\mathbb{R}$ *that have compact support, and let T be a linear transformation of D into the linear space M of all complex-valued measurable functions on* $\mathbb{R}$. *Let* $1 \leq p_0 < p_1 < \infty$ *be given, and suppose that:*

(1) *There exist numbers* q_0 *and* M_0, *with* $1 < q_0 \leq \infty$, *such that* $\|T(f)\|_{q_0} \leq M_0\|f\|_{p_0}$ *for all* $f \in D$; *i.e., T has a unique extension to a bounded operator* T_0 *from* L^{p_0} *into* L^{q_0}, *and* $\|T_0\| \leq M_0$.

(2) *There exist numbers* q_1 *and* M_1, *with* $1 < q_1 \leq \infty$, *such that* $\|T(f)\|_{q_1} \leq M_1\|f\|_{p_1}$ *for all* $f \in D$; *i.e.,* T *has a unique extension to a bounded operator* T_1 *from* L^{p_1} *into* L^{q_1}, *and* $\|T_1\| \leq M_1$.

Let p satisfy $p_0 < p < p_1$, and define $t \in (0,1)$ by

$$1/p = (1-t)/p_0 + t/p_1;$$

i.e.,

$$t = \frac{1/p - 1/p_0}{1/p_1 - 1/p_0}.$$

Now define q by

$$1/q = (1-t)/q_0 + t/q_1.$$

Then

$$\|T(f)\|_q \leq M_p\|f\|_p,$$

for all $f \in D$, where

$$M_p = M_0^{1-t}M_1^t.$$

Hence, T has a unique extension to a bounded operator T_p from L^p into L^q, and $\|T_p\| \leq M_p$.

PROOF. For any $1 < r < \infty$, we write r' for the conjugate number defined by $1/r + 1/r' = 1$. Let $f \in D$ be given, and suppose that $\|f\|_p = 1$. If the theorem holds for all such f, it will hold for all $f \in D$. (Why?) Because $T(f)$ belongs to L^{q_0} and to L^{q_1} by hypothesis, it follows that $T(f) \in L^q$, so that it is only the inequality on the norms that we must verify. We will show that $|\int [T(f)](y)g(y)\,dy| \leq m_p$, whenever $g \in D \cap L^{q'}$ with $\|g\|_{q'} = 1$. This will complete the proof (see Exercise 4.13). Thus, let g be such a function. Write $f = \sum_{j=1}^n a_j \chi_{A_j}$ and $g = \sum_{k=1}^m b_k \chi_{B_k}$, for $\{A_j\}$ and $\{B_k\}$ disjoint bounded measurable sets and a_j and b_k nonzero complex numbers.

For each $z \in \mathbb{C}$, define

$$\alpha(z) = (1-z)/p_0 + z/p_1$$

and

$$\beta(z) = (1-z)/q_0' + z/q_1'.$$

Note that $\alpha(t) = 1/p$ and $\beta(t) = 1/q'$.

We extend the definition of the signum function to the complex plane as follows: If λ is a nonzero complex number, define $\text{sgn}(\lambda)$ to be $\lambda/|\lambda|$. For each complex z, define the simple functions

$$f_z = \sum_{j=1}^{n} \text{sgn}(a_j)|a_j|^{\alpha(z)/\alpha(t)}\chi_{A_j}$$

and

$$g_z = \sum_{k=1}^{m} \text{sgn}(b_k)|b_k|^{\beta(z)/\beta(t)}\chi_{B_k},$$

and finally put

$$F(z) = \int [T(f_z)](y)g_z(y)\,dy$$

$$= \sum_{j=1}^{n}\sum_{k=1}^{m} \text{sgn}(a_j)\text{sgn}(b_k)|a_j|^{p\alpha(z)}|b_k|^{q'\beta(z)} \int [T(\chi_{A_j})](y)\chi_{B_k}(y)\,dy$$

$$= \sum_{j=1}^{n}\sum_{k=1}^{m} c_{jk}e^{d_{jk}z},$$

where the c_{jk}'s are complex numbers and the d_{jk}'s are real numbers.

Observe that F is an entire function of the complex variable z, and that it is bounded on the closed strip $0 \leq \Re z \leq 1$. Note also that $\int [T(f)](y)g(y)\,dy$, the quantity we wish to estimate, is precisely $F(t)$.

Observe next that

$$
\sup_{s\in\mathbb{R}} |F(is)| = \sup_{s} |\int [T(f_{is})](y)g_{is}(y)\,dy|
$$

$$
\leq \sup_{s} (\int |[T(f_{is})](y)|^{q_0}\,dy)^{1/q_0} (\int |g_{is}(y)|^{q_0'}\,dy)^{1/q_0'}
$$

$$
\leq \sup_{s} m_0 \|f_{is}\|_{p_0} \|g_{is}\|_{q_0'}
$$

$$
= \sup_{s} m_0 (\int \sum_{j} |(|a_j|^{p_0\alpha(is)/\alpha(t)})|\chi_{A_j}(y)\,dy)^{1/p_0}
$$

$$
\times (\int \sum_{k} |(|b_k|^{q_0'\beta(is)/\beta(t)})|\chi_{B_k}(y)\,dy)^{1/q_0'}
$$

$$
= m_0 \sup_{s} \int \sum_{j} |a_j|^p \chi_{A_j}(y)\,dy)^{1/p_0}
$$

$$
\times (\int \sum_{k} |b_k|^{q'} \chi_{B_k}(y)\,dy)^{1/q_0'}
$$

$$
= m_0 \|f\|_p^{p/p_0} \|g\|_{q'}^{q'/q_0'}
$$

$$
= m_0.
$$

By a similar calculation, we see that

$$
\sup_{s\in\mathbb{R}} |F(1+is)| \leq m_1.
$$

The proof of the theorem is then completed by appealing to the lemma from complex variables that follows.

LEMMA. *Suppose F is a complex-valued function that is bounded and continuous on the closed strip $0 \leq \Re z \leq 1$ and analytic on the open strip $0 < \Re z < 1$. Assume that m_0 and m_1 are real numbers satisfying*

$$
m_0 \geq \sup_{s\in\mathbb{R}} |F(is)|
$$

and

$$
m_1 \geq \sup_{s\in\mathbb{R}} |F(1+is)|.
$$

Then

$$
\sup_{s\in\mathbb{R}} |F(t+is)| \leq m_0^{1-t} m_1^t
$$

for all $0 \le t \le 1$.

PROOF. We may assume that m_0 and m_1 are positive. Define a function G on the strip $0 \le \Re z \le 1$ by

$$G(z) = F(z)/m_0^{1-z} m_1^z.$$

Then G is continuous and bounded on this strip and is analytic on the open strip $0 < \Re z < 1$. It will suffice to prove that

$$\sup_{s \in \mathbb{R}} |G(t + is)| \le 1.$$

For each positive integer n, define $G_n(z) = G(z)e^{z^2/n}$. Then each function G_n is continuous and bounded on the strip $0 \le \Re z \le 1$ and analytic on the open strip $0 < \Re z < 1$. Also, $G(z) = \lim G_n(z)$ for all z in the strip. It will suffice then to show that $\lim |G_n(z)| \le 1$ for each z for which $0 < \Re z < 1$. Fix $z_0 = x_0 + iy_0$ in the open strip, and choose a $Y > |y_0|$ such that $|G_n(z)| = |G_n(x + iy)| = |G(z)|e^{(x^2-y^2)/n} < 1$ whenever $|y| \ge Y$. Let Γ be the rectangular contour determined by the four points $(0, -Y)$, $(1, -Y)$, $(1, Y)$, and $(0, Y)$. Then, by the Maximum Modulus Theorem, we have

$$
\begin{aligned}
|G_n(z_0)| &\le \max_{z \in \Gamma} |G_n(z)| \\
&\le \max(1, \sup_{s \in \mathbb{R}} |G_n(1 + is)| \ , \ 1, \sup_{s \in \mathbb{R}} |G_n(is)|) \\
&= e^{1/n},
\end{aligned}
$$

proving that $\lim |G_n(z_0)| \le 1$, and this completes the proof of the lemma.

EXERCISE 4.15. Verify that the Riesz Interpolation Theorem holds with $\mathbb{R}$ replaced by any regular σ-finite measure space.

CHAPTER V

DUAL SPACES

DEFINITION Let $(X, \mathcal{T})$ be a (real) locally convex topological vector space. By the *dual space* X^*, or $(X, \mathcal{T})^*$, of X we mean the set of all continuous linear functionals on X.

By the *weak topology* on X we mean the weakest topology $\mathcal{W}$ on X for which each $f \in X^*$ is continuous. In this context, the topology $\mathcal{T}$ is called the *strong topology* or *original topology* on X.

EXERCISE 5.1. (a) Prove that X^* is a vector space under pointwise operations.

(b) Show that $\mathcal{W} \subseteq \mathcal{T}$. Show also that $(X, \mathcal{W})$ is a locally convex topological vector space.

(c) Show that if X is infinite dimensional then every weak neighborhood of 0 contains a nontrivial subspace M of X. HINT: If $V = \cap_{i=1}^n f_i^{-1}(U_i)$, and if $M = \cap_{i=1}^n \ker(f_i)$, then $M \subseteq V$.

(d) Show that a linear functional f on X is strongly continuous if and only if it is weakly continuous; i.e., prove that $(X, \mathcal{T})^* = (X, \mathcal{W})^*$.

(e) Prove that X is finite dimensional if and only if X^* is finite dimensional, in which case X and X^* have the same dimension.

EXERCISE 5.2. (a) For $1 < p < \infty$, let X be the normed linear space $L^p(\mathbb{R})$. For each $g \in L^{p'}(\mathbb{R})$ $(1/p + 1/p' = 1)$, define a linear functional ϕ_g on X by

$$\phi_g(f) = \int f(x)g(x)\, dx.$$

Prove that the map $g \to \phi_g$ is a vector space isomorphism of $L^{p'}(\mathbb{R})$ onto X^*.

(b) By analogy to part a, show that $L^\infty(\mathbb{R})$ is isomorphic as a vector space to $L^1(\mathbb{R})^*$.

(c) Let c_0 be the normed linear space of real sequences $\{a_0, a_1, \ldots\}$ for which $\lim a_n = 0$ with respect to the norm defined by $\|\{a_n\}\| = \max |a_n|$. Show that c_0^* is algebraically isomorphic to l^1, where l^1 is the linear space of all absolutely summable sequences $\{b_0, b_1, \ldots\}$. HINT: If $f \in c_0^*$, define b_n to be $f(e^n)$, where e^n is the element of c_0 that is 1 in the nth position and 0 elsewhere.

(d) In each of parts a through c, show that the weak and strong topologies are different. Exhibit, in fact, nets (sequences) which converge weakly but not strongly.

(e) Let $X = L^\infty(\mathbb{R})$. For each function $g \in L^1(\mathbb{R})$, define ϕ_g on X by $\phi_g(f) = \int fg$. Show that ϕ_g is an element of X^*. Similarly, for each finite Borel measure μ on $\mathbb{R}$, define ϕ_μ on X by $\phi_\mu(f) = \int f \, d\mu$. Show that ϕ_μ is an element of X^*. Conclude that, in this sense, $L^1(\mathbb{R})$ is a proper subset of $(L^\infty)^*$.

(f) Let Δ be a second countable locally compact Hausdorff space, and let X be the normed linear space $C_0(\Delta)$ equipped with the supremum norm. Identify X^*.

(g) Let $X_1, \ldots, X_n$ be locally convex topological vector spaces. If $X = \bigoplus_{i=1}^n X_i$, show that X^* is isomorphic to $\bigoplus_{i=1}^n X_i^*$.

THEOREM 5.1. (Relation between the Weak and Strong Topologies) Let $(X, \mathcal{T})$ be a locally convex topological vector space.

(1) Let A be a convex subset of X. Then A is strongly closed if and only if it is weakly closed.

(2) If A is a convex subset of X, then the weak closure of A equals the strong closure of A.

(3) If $\{x_\alpha\}$ is a net in X that converges weakly to an element x, then there exists a net $\{y_\beta\}$, for which each y_β is a (finite) convex combination of some of the x_α's, such that $\{y_\beta\}$ converges strongly to x. If $\mathcal{T}$ is metrizable, then the net $\{y_\beta\}$ can be chosen to be a sequence.

PROOF. If A is a weakly closed subset, then it is strongly closed since $\mathcal{W} \subseteq \mathcal{T}$. Conversely, suppose that A is a strongly closed convex set and let $x \in X$ be an element not in A. Then, by the Separation Theorem, there exists a continuous linear functional ϕ on X, and a real number s, such that $\phi(y) \leq s$ for all $y \in A$ and $\phi(x) > s$. But then the set $\phi^{-1}(s,\infty)$ is a weakly open subset of X that contains x and is disjoint from A, proving that A is weakly closed, as desired.

If A is a convex subset of X, and if B is the weak closure and C is the strong closure, then clearly $A \subseteq C \subseteq B$. On the other hand, C is convex and strongly closed, hence C is weakly closed. Therefore, $B = C$, and part 2 is proved.

Now let $\{x_\alpha\}$ be a weakly convergent net in X, and let A be the convex hull of the x_α's. If $x = \lim_W x_\alpha$, then x belongs to the weak closure of A, whence to the strong closure of A. Let $\{y_\beta\}$ be a net (sequence if $\mathcal{T}$ is metrizable) of elements of A that converges strongly to x. Then each y_β is a finite convex combination of certain of the x_α's, and part 3 is proved.

DEFINITION. Let X be a locally convex topological vector space, and let X^* be its dual space. For each $x \in X$, define a function $\hat{x}$ on X^* by $\hat{x}(f) = f(x)$. By the *weak* * *topology* on X^*, we mean the weakest topology $\mathcal{W}^*$ on X^* for which each function $\hat{x}$, for $x \in X$, is continuous.

THEOREM 5.2. (Duality Theorem) *Let $(X, \mathcal{T})$ be a locally convex topological vector space, and let X^* be its dual space. Then:*

(1) *Each function $\hat{x}$ is a linear functional on X^*.*

(2) *$(X^*, \mathcal{W}^*)$ is a locally convex topological vector space. (Each $\hat{x}$ is continuous on $(X^*, \mathcal{W}^*)$.)*

(3) *If ϕ is a continuous linear functional on $(X^*, \mathcal{W}^*)$, then there exists an $x \in X$ such that $\phi = \hat{x}$; i.e., the map $x \to \hat{x}$ is a linear transformation of X onto $(X^*, \mathcal{W}^*)^*$.*

(4) *The map $x \to \hat{x}$ is a topological isomorphism between $(X, \mathcal{W})$ and $((X^*, \mathcal{W}^*)^*, \mathcal{W}^*)$.*

PROOF. If $x \in X$, then

$$\hat{x}(af + bg) = (af + bg)(x)$$
$$= af(x) + bg(x)$$
$$= a\hat{x}(f) + b\hat{x}(g),$$

proving part 1.

By the definition of the topology $\mathcal{W}^*$, we see that each $\hat{x}$ is continuous. Also, the set of all functions $\{\hat{x}\}$ separate the points of X^*, for if $f, g \in X^*$, with $f \neq g$, then $f - g$ is not the 0 functional. Hence there exists an $x \in X$ for which $(f - g)(x)$, which is $\hat{x}(f) - \hat{x}(g)$, is not 0. Therefore, the weak topology on X^*, generated by the $\hat{x}$'s, is a locally convex topology. See part c of Exercise 3.11.

Now suppose ϕ is a continuous linear functional on $(X^*, \mathcal{W}^*)$, and let M be the kernel of ϕ. If $M = X^*$, then ϕ is the 0 functional, which is $\hat{0}$. Assume then that there exists an $f \in X^*$, for which $\phi(f) = 1$, whence $f \notin M$. Since ϕ is continuous, M is a closed subset in X^*, and there exists a weak* neighborhood V of f which is disjoint from M. Therefore, by the definition of the topology $\mathcal{W}^*$, there exists a finite set $x_1, \ldots, x_n$ of elements of X and a finite set $\epsilon_1, \ldots, \epsilon_n$ of positive real numbers such that

$$V = \{g \in X^* : |\hat{x}_i(g) - \hat{x}_i(f)| < \epsilon_i, \ 1 \leq i \leq n\}.$$

Define a map $R : X^* \to \mathbb{R}^n$ by

$$R(g) = (\hat{x}_1(g), \ldots, \hat{x}_n(g)).$$

Clearly R is a continuous linear transformation of X^* into $\mathbb{R}^n$. Now $R(f) \notin R(M)$, for otherwise there would exist a $g \in M$ such that $\hat{x}_i(g) = \hat{x}_i(f)$ for all i. But this would imply that $g \in V \cap M$, contradicting the choice of the neighborhood V. Also, $R(M)$ is a subspace of $\mathbb{R}^n$, so contains 0, implying then that $R(f) \neq 0$. Suppose $R(M)$ is of dimension $j < n$. Let $\alpha_1, \ldots, \alpha_n$ be a basis for $\mathbb{R}^n$, such that $\alpha_1 = R(f)$ and $\alpha_i \in R(M)$ for $2 \leq i \leq j + 1$. We define a linear functional p on $\mathbb{R}^n$ by setting $p(\alpha_1) = 1$ and $p(\alpha_i) = 0$ for $2 \leq i \leq n$.

Now, $p \circ R$ is a continuous linear functional on X^*. If $g \in M$, then $(p \circ R)(g) = p(R(g)) = 0$, since $R(g) \in R(M)$, which is in the span of $\alpha_2, \ldots, \alpha_n$. Also, $(p \circ R)(f) = p(R(f)) = 1$, since $R(f) = \alpha_1$. So, $p \circ R$ is a linear functional on X^* which has the same kernel M as ϕ and agrees with ϕ on f. Therefore, $\phi - p \circ R = 0$ everywhere, and $\phi = p \circ R$.

Let $e_1, \ldots, e_n$ denote the standard basis for $\mathbb{R}^n$, and let A be the $n \times n$ matrix relating the bases $e_1, \ldots, e_n$ and $\alpha_1, \ldots, \alpha_n$. That is, $e_i = \sum_{j=1}^n A_{ij}\alpha_j$. Then, if $\alpha = (a_1, \ldots, a_n) = \sum_{i=1}^n a_i e_i$, we have

$$p(\alpha) = \sum_{i=1}^n a_i p(e_i)$$

$$= \sum_{i=1}^n a_i \sum_{j=1}^n A_{ij} p(\alpha_j)$$

$$= \sum_{i=1}^n a_i A_{i1}.$$

Therefore,

$$\phi(g) = p \circ R(g)$$
$$= p(R(g))$$
$$= p((\widehat{x_1}(g), \ldots, \widehat{x_n}(g)))$$
$$= \sum_{i=1}^{n} A_{i1} \widehat{x_i}(g)$$
$$= (\sum_{i=1}^{n} A_{i1} \widehat{x_i})(g)$$
$$= \overbrace{\sum_{i=1}^{n} A_{i1} x_i}(g)$$
$$= \hat{x}(g),$$

where $x = \sum_{i=1}^{n} A_{i1} x_i$, and this proves part 3.

We leave the proof of part 4 to the exercises.

EXERCISE 5.3. Prove part 4 of the preceding theorem. HINT: Show that a net $\{x_\alpha\}$ converges in the weak topology of X to an element x if and only if the net $\{\widehat{x_\alpha}\}$ converges in the weak* topology of $(X^*, W^*)^*$ to the element $\hat{x}$.

DEFINITION. If T is a continuous linear transformation from a locally convex topological vector space X into a locally convex topological vector space Y, we define the transpose T^* of T to be the function from Y^* into X^* given by

$$[T^*(f)](x) = f(T(x)).$$

EXERCISE 5.4. If T is a continuous linear transformation from a locally convex topological vector space X into a locally convex topological vector space Y, show that the transpose T^* is a continuous linear transformation from (Y^*, W^*) into (X^*, W^*).

EXERCISE 5.5. (Continuous Linear Functionals on Dense Subspaces, Part 1) Let X be a locally convex topological vector space, and let Y be a dense subspace of X.

(a) Prove that Y is a locally convex topological vector space in the relative topology.

(b) Let f be a continuous linear functional on Y, and let x be an element of X that is not in Y. Let $\{y_\alpha\}$ be a net of elements of Y for which $x = \lim y_\alpha$. Prove that the net $\{f(y_\alpha)\}$ is a Cauchy net in $\mathbb{R}$, and hence converges.

(c) Let f be in Y^*. Show that f has a unique extension to a continuous linear functional f' on X. HINT: Show that f' is well-defined and is bounded on a neighborhood of 0.

(d) Conclude that the map $f \to f'$ of part c is an isomorphism of the vector space Y^* onto the vector space X^*. Compare with Exercise 4.11, part a.

EXERCISE 5.6. (Continuous Linear Functionals on Dense Subspaces, Part 2) Let Y be a dense subspace of a locally convex topological vector space X, and equip Y with the relative topology. If Y is a proper subspace of X, show that the map $f \to f'$ of part c of the preceding exercise is not a topological isomorphism of $(Y^*, \mathcal{W}^*)$ and $(X^*, \mathcal{W}^*)$. HINT: Use Theorem 5.2. Compare with part b of Exercise 4.11.

EXERCISE 5.7. (Weak Topologies and Metrizability) Let X be a locally convex topological vector space, and let X^* be its dual space.

(a) Show that the weak topology on X is the weakest topology for which each f_α is continuous, where the f_α's form a basis for the vector space X^*. Similarly, show that the weak* topology on X^* is the weakest topology for which each $\widehat{x_\alpha}$ is continuous, where the x_α's form a basis for the vector space X.

(b) Show that the weak* topology on X^* is metrizable if and only if, as a vector space, X has a countable basis. Show also that the weak topology on X is metrizable if and only if, as a vector space, X^* has a countable basis.

(c) Let X be the locally convex topological vector space $\prod_{n=1}^\infty \mathbb{R}$. Compute X^*, and verify that it has a countable basis. HINT: Show that X^* can be identified with the space of sequences $\{a_1, a_2, \dots\}$ that are eventually 0. That is, as a vector space, X^* is isomorphic to c_c.

(d) Deduce that the topological vector space $X = \prod_{n=1}^\infty \mathbb{R}$ is a Frechet space that is not normable.

DEFINITION. Let $\mathcal{S}$ be Schwartz space, i.e., the countably normed vector space of Exercise 3.10. Elements of the dual space $\mathcal{S}^*$ of $\mathcal{S}$ are called *tempered distributions* on $\mathbb{R}$.

EXERCISE 5.8. (Properties of Tempered Distributions)

(a) If h is a measurable function on $\mathbb{R}$, for which there exists a positive integer n such that $h(x)/(1 + |x|^n)$ is in L^1, we say that h is a *tempered function*. If h is a tempered function, show that the assignment $f \to$

$\int_{-\infty}^{\infty} h(t)f(t)\, dt$ is a tempered distribution u_h. Show further that h is integrable over any finite interval and that the function k, defined by $k(x) = \int_0^x h(t)\, dt$ if $x \geq 0$, and by $k(x) = -\int_x^0 h(t)\, dt$ if $x \leq 0$, also is a tempered function.

(b) Show that $h(x) = 1/x$ is not a tempered function but that the assignment

$$f \to \lim_{\delta \to 0} \int_{|t| \geq \delta} (1/t) f(t)\, dt$$

is a tempered distribution. (Integrate by parts and use the mean value theorem.) Show further that $h(x) = 1/x^2$ is not a tempered function, and also that the assignment

$$f \to \lim_{\delta \to 0} \int_{|t| \geq \delta} (1/t^2) f(t)\, dt$$

is not a tempered distribution. (In fact, this limit fails to exist in general.) In some sense, then, $1/x$ can be considered to determine a tempered distribution but $1/x^2$ cannot.

(c) If μ is a Borel measure on $\mathbb{R}$, for which there exists a positive integer n such that $\int (1/(1 + |x|^n))\, d\mu(x)$ is finite, we say that μ is a tempered measure. If μ is a tempered measure, show that the assignment $f \to \int_{-\infty}^{\infty} f(t)\, d\mu(t)$ is a tempered distribution u_μ.

(d) Show that the linear functional δ, defined on $\mathcal{S}$ by $\delta(f) = f(0)$ (the so-called Dirac δ-function), is a tempered distribution, and show that $\delta = u_\mu$ for some tempered measure μ.

(e) Show that the linear functional δ', defined on $\mathcal{S}$ by $\delta'(f) = -f'(0)$, is a tempered distribution, and show that δ' is not the same as any tempered distribution of the form u_h or u_μ. HINT: Show that δ' fails to satisfy the dominated convergence theorem.

(f) Let u be a tempered distribution. Define a linear functional u' on $\mathcal{S}$ by $u'(f) = -u(f')$. Prove that u' is a tempered distribution. We call u' the distributional derivative of u. As usual, we write $u^{(n)}$ for the nth distributional derivative of u. We have that

$$u^{(n)}(f) = (-1)^n U(f^{(n)}).$$

Verify that if h is a C^∞ function on $\mathbb{R}$, for which both h and h' are tempered functions, then the distributional derivative $(u_h)'$ of u_h coincides with the tempered distribution $u_{h'}$, showing that distributional differentiation is a generalization of ordinary differentiation. Explain why the minus sign is present in the definition of the distributional derivative.

(g) If h is defined by $h(x) = \ln(|x|)$, show that h is a tempered function, that h' is not a tempered function, but that

$$(u_h)'(f) = \lim_{\delta \to 0} \int_{|t| \geq \delta} (1/t) f(t)\, dt = \lim_{\delta \to 0} \int_{|t| \geq \delta} h'(t) f(t)\, dt.$$

Moreover, compute $(u_h)^{(2)}$ and show that it cannot be interpreted in any way as being integration against a function.

(h) If h is a tempered function, show that there exists a tempered function k whose distributional derivative is h, i.e., $u'_k = u_h$.

(i) Suppose h is a tempered function for which the distributional derivative u'_h of the tempered distribution u_h is 0. Prove that there exists a constant c such that $h(x) = c$ for almost all x. HINT: Verify and use the fact that, if $\int_a^b h(x) f(x)\, dx = 0$ for all functions f that satisfy $\int_a^b f(x)\, dx = 0$, then h agrees with a constant function almost everywhere on $[a, b]$.

THEOREM 5.3. (Representing a Tempered Distribution as the Derivative of a Function) *Let u be a tempered distribution. Then there exists a tempered function h and a nonnegative integer N such that u is the Nth distributional derivative $u_h^{(N)}$ of the tempered distribution u_h. We say then that every tempered distribution is the nth derivative of a tempered function.*

PROOF. Let $u \in S^*$ be given. Recall that S is a countably normed space, where the norms $\{\rho_n\}$ are defined by

$$\rho_n(f) = \sup_x \max_{0 \leq i, j \leq n} |x^j f^{(i)}(x)|.$$

We see then that $\rho_n \leq \rho_{n+1}$ for all n. Therefore, according to part e of Exercise 3.8, there exists an integer N and a constant M such that $|u(f)| \leq M\rho_N(f)$ for all $f \in S$. Now, for each $f \in S$, and each nonnegative integer n, set

$$p_n(f) = \max_{0 \leq i, j \leq n} \int_{-\infty}^{\infty} |t^j f^{(i)}(t)|\, dt.$$

There exists a point x_0 and integers i_0 and j_0 such that

$$\rho_N(f) = |x_0^{j_0} f^{(i_0)}(x_0)|$$

$$= \left| \int_{-\infty}^{x_0} j_0 t^{j_0 - 1} f^{(i_0)}(t) + t^{j_0} f^{(i_0 + 1)}(t)\, dt \right|$$

$$\leq \int_{-\infty}^{\infty} j_0 |t^{j_0 - 1} f^{(i_0)}(t) + t^{j_0} f^{(i_0 + 1)}(t)|\, dt$$

$$\leq (N + 1) p_{N+1}(f),$$

showing that $|u(f)| \le M(N+1)p_{N+1}(f)$ for all $f \in \mathcal{S}$.

Let Y be the normed linear space $\mathcal{S}$, equipped with the norm p_{N+1}. Let

$$X = \bigoplus_{i,j=0}^{N+1} L^1(\mathbb{R}),$$

and define a map $F : Y \to X$ by

$$[F(f)]_{i,j}(x) = x^j f^{(i)}(x).$$

Then, using the max norm on the direct sum space X, we see that F is a linear isometry of Y into X. Moreover, the tempered distribution u is a continuous linear functional on Y and hence determines a continuous linear functional $\tilde{u}$ on the subspace $F(Y)$ of X. By the Hahn-Banach Theorem, there exists a continuous linear functional ϕ on X whose restriction to $F(Y)$ coincides with $\tilde{u}$.

Now $X^* = \bigoplus_{i,j=0}^{N+1} L^\infty(\mathbb{R})$, whence there exist L^∞ functions $v_{i,j}$ such that

$$\phi(g) = \sum_{i=0}^{N+1} \sum_{j=0}^{N+1} \int g_{i,j}(t) v_{i,j}(t)\, dt$$

for all $g = \{g_{i,j}\} \in X$. Hence, for $f \in \mathcal{S}$, we have

$$
\begin{aligned}
u(f) &= \tilde{u}(F(f)) \\
&= \phi(F(f)) \\
&= \sum_{i=0}^{N+1} \sum_{j=0}^{N+1} \int [F(f)]_{i,j}(t) v_{i,j}(t)\, dt \\
&= \sum_{i=0}^{N+1} \sum_{j=0}^{N+1} \int t^j f^{(i)}(t) v_{i,j}(t)\, dt \\
&= \sum_{i=0}^{N+1} \int f^{(i)}(t) \Big(\sum_{j=0}^{N+1} t^j v_{i,j}(t) \Big)\, dt \\
&= \sum_{i=0}^{N+1} \int f^{(i)}(t) v_i(t)\, dt,
\end{aligned}
$$

where $v_i(t) = \sum_{j=0}^{N+1} t^j v_{i,j}(t)$. Clearly, each v_i is a tempered function, and we let w_i be a tempered function whose $(N+1-i)$th distributional derivative is v_i. (See part h of Exercise 5.8.) Then,

$$u(f) = \sum_{i=0}^{N+1} \int f^{(i)}(t) w_i^{(N+1-i)}(t) \, dt$$

$$= \sum_{i=0}^{N+1} (-1)^{N+1-i} \int f^{(N+1)}(t) w_i(t) \, dt$$

$$= \int f^{(N+1)}(t) w(t) \, dt,$$

where $w(t) = \sum_{i=0}^{N+1} (-1)^{N+1-i} w_i$. Hence $u(f) = \int f^{(N+1)} w$, or

$$u = (-1)^{N+1} u_w^{(N+1)} = u_h^{N+1},$$

where $h = (-1)^{N+1} w$, and this completes the proof.

DEFINITION. If f is a continuous linear functional on a normed linear space X, define the *norm* $\|f\|$ of f as in Chapter IV by

$$\|f\| = \sup_{\substack{x \in X \\ \|x\| \leq 1}} |f(x)|.$$

DEFINITION. If X is a normed linear space, we define the *conjugate space* of X to be the dual space X^* of X equipped with the norm defined above.

EXERCISE 5.9. Let X be a normed linear space, and let X^* be its dual space. Denote by $\mathcal{W}^*$ the weak* topology on X^* and by $\mathcal{N}$ the topology on X^* defined by the norm.
(a) Show that the conjugate space X^* of X is a Banach space.
(b) Show that, if X is infinite dimensional, then the weak topology on X is different from the norm topology on X and that the weak* topology on the dual space X^* is different from the norm topology on X^*. HINT: Use part c of Exercise 5.1. Note then that the two dual spaces $(X^*, \mathcal{W}^*)^*$ and $(X^*, \mathcal{N})^* \equiv X^{**}$ may be different.

EXERCISE 5.10. (a) Show that the vector space isomorphisms of parts a through c of Exercise 5.2 are isometric isomorphisms.
(b) Let X be a normed linear space and let X^* denote its conjugate space. Let X^{**} denote the conjugate space of the normed linear space X^*. If $x \in X$, define $\hat{x}$ on X^* by $\hat{x}(f) = f(x)$. Show that $\hat{x} \in X^{**}$.

(c) Again let X be a normed linear space and let X^* denote its conjugate space. Prove that $(X^*, W^*)^* \subseteq X^{**}$; i.e., show that every continuous linear functional on (X^*, W^*) is continuous with respect to the norm topology on X^*.

(d) Let the notation be as in part b. Prove that the map $x \to \hat{x}$ is continuous from (X, W) into (X^{**}, W^*).

THEOREM 5.4. *Let X be a normed linear space.*

(1) *If Y is a dense subspace of X, then the restriction map $g \to \tilde{g}$ of X^* into Y^* is an isometric isomorphism of X^* onto Y^*.*

(2) *The weak* topology W^* on X^* is weaker than the topology defined by the norm on X^*.*

(3) *The map $x \to \hat{x}$ is an isometric isomorphism of the normed linear space X into the conjugate space X^{**} of the normed linear space X^*.*

PROOF. That the restriction map $g \to \tilde{g}$ is an isometric isomorphism of X^* onto Y^* follows from part c of Exercise 5.5 and the definitions of the norms.

If $x \in X$, then $\hat{x}$ is a linear functional on X^* and $|\hat{x}(f)| = |f(x)| \leq \|x\|\|f\|$, showing that $\hat{x}$ is a continuous linear functional in the norm topology of the Banach space X^*, and that $\|\hat{x}\| \leq \|x\|$. Since the weak* topology is the weakest making each $\hat{x}$ continuous, it follows that W^* is contained in the norm topology on X^*.

Finally, given an $x \in X$, there exists by the Hahn-Banach Theorem an $f \in X^*$ for which $\|f\| = 1$ and $f(x) = \|x\|$. Therefore, $\hat{x}(f) = \|x\|$, showing that $\|\hat{x}\| \geq \|x\|$, and the proof is complete.

EXERCISE 5.11. (The Normed Linear Space of Finite Complex Measures on a Second Countable Locally Compact Hausdorff Space, Part 1) Let Δ be a second countable locally compact Hausdorff space, and let $M(\Delta)$ be the complex vector space of all finite complex Borel measures on Δ. Recall that a finite complex Borel measure on Δ is a map μ of the σ-algebra $\mathcal{B}$ of Borel subsets of Δ into $\mathbb{C}$ satisfying:

(1) There exists a positive K such that $|\mu(E)| \leq K$ for all $E \in \mathcal{B}$.

(2) $\mu(\emptyset) = 0$.

(3) If $\{E_n\}$ is a sequence of pairwise disjoint Borel sets, then the series $\sum \mu(E_n)$ is absolutely summable and $\mu(\cup E_n) = \sum \mu(E_n)$.

(a) If $\mu \in M(\Delta)$, show that μ can be written uniquely as $\mu = \mu_1 + i\mu_2$, where μ_1 and μ_2 are finite Borel signed measures on Δ. Show further that each μ_i may be written uniquely as $\mu_i = \mu_{i1} - \mu_{i2}$, where each μ_{ij}

is a finite positive Borel measure, and where μ_{i1} and μ_{i2} are mutually singular.

(b) Let $M(\Delta)$ be as in part a, and let μ be an element of $M(\Delta)$. Given a Borel set E and an $\epsilon > 0$, show that there exists a compact set K and an open set U for which $K \subseteq E \subseteq U$ such that $|\mu(U - K)| < \epsilon$. HINT: Use the fact that Δ is σ-compact, and consider the collection of sets E for which the desired condition holds. Show that this is a σ-algebra that contains the open sets.

EXERCISE 5.12. (The Normed Linear Space of Finite Complex Measures on a Second Countable Locally Compact Hausdorff Space, Part 2) Let $M(\Delta)$ be as in the previous exercise, and for each $\mu \in M(\Delta)$ define

$$\|\mu\| = \sup \sum_{i=1}^{n} |\mu(E_i)|,$$

where the supremum is taken over all partitions $E_1, \ldots, E_n$ of Δ into a finite union of pairwise disjoint Borel subsets.

(a) Show that $\|\mu\| < \infty$.

(b) Prove that $M(\Delta)$ is a normed linear space with respect to the above definition of $\|\mu\|$. This norm is called the *total variation norm*.

(c) If h is a bounded, complex-valued Borel function on Δ, and if $\mu \in M(\Delta)$, show that

$$\left| \int h \, d\mu \right| \leq \|h\|_\infty \|\mu\|.$$

HINT: Do this first for simple functions.

(d) For each $\mu \in M(\Delta)$, define a linear functional ϕ_μ on the complex Banach space $C_0(\Delta)$ by

$$\phi_\mu(f) = \int f \, d\mu.$$

Prove that the map $\mu \to \phi_\mu$ is a norm-decreasing isomorphism of the normed linear space $M(\Delta)$ onto $C_0(\Delta)^*$.

(e) Let μ be an element of $M(\Delta)$. Prove that

$$\|\mu\| = \sup \sum_{i=1}^{n} |\mu(K_i)|,$$

where the supremum is taken over all n-tuples $K_1, \dots, K_n$ of pairwise disjoint compact subsets of Δ.

(f) Let μ be an element of $M(\Delta)$, and let ϕ_μ be the element of $C_0(\Delta)^*$ defined in part d. Prove that $\|\phi_\mu\| = \|\mu\|$. Conclude that $M(\Delta)$ is a Banach space with respect to the norm $\|\mu\|$ and that it is isometrically isomorphic to $C_0(\Delta)^*$.

EXERCISE 5.13. Let X be the normed linear space c_0. See part c of Exercise 5.2.

(a) Compute the conjugate space c_0^*.

(b) Compute c_0^{**} and $(C_0^*, \mathcal{W}^*)^*$. Conclude that $(X^*, \mathcal{W}^*)^*$ can be properly contained in X^{**}; i.e., there can exist linear functionals on X^* that are continuous with respect to the norm topology but not continuous with respect to the weak* topology.

DEFINITION. A Banach space X is called *reflexive* if the map $x \to \hat{x}$ is an (isometric) isomorphism of X onto X^{**}. In general, X^{**} is called the *second dual* or *second conjugate* of X.

EXERCISE 5.14. (Relation among the Weak, Weak*, and Norm Topologies) Let X be a normed linear space. Let $\mathcal{N}$ denote the topology on X^* determined by the norm, let $\mathcal{W}$ denote the weak topology on the locally convex topological vector space $(X^*, \mathcal{N})$, and let $\mathcal{W}^*$ denote the weak* topology on X^*.

(a) If X is finite dimensional, show that all three topologies are the same.

(b) If X is an infinite-dimensional reflexive Banach space, show that $\mathcal{W}^* = \mathcal{W}$, and that $\mathcal{W} \subset \mathcal{N}$.

(c) If X is not reflexive, show that $\mathcal{W}^* \subseteq \mathcal{W} \subset \mathcal{N}$.

(d) Let X be a nonreflexive Banach space. Show that there exists a subspace of X^* which is closed in the norm topology $\mathcal{N}$ (whence also in the weak topology $\mathcal{W}$) but not closed in the weak* topology $\mathcal{W}^*$, and conclude then that $\mathcal{W}^* \subset \mathcal{W}$. HINT: Let ϕ be a norm continuous linear functional that is not weak* continuous, and examine its kernel.

(e) Suppose X is an infinite dimensional Banach space. Prove that neither $(X, \mathcal{W})$ nor $(X^*, \mathcal{W}^*)$ is metrizable. HINT: Use the Baire Category Theorem to show that any Banach space having a countable basis must be finite dimensional.

DEFINITION. Let X and Y be normed linear spaces, and let T be a continuous linear transformation from X into Y. The transpose T^* of T is called the *adjoint* of T when it is regarded as a linear transformation from the normed linear space Y^* into the normed linear space X^*.

THEOREM 5.5. Let T be a continuous linear transformation from a normed linear space X into a normed linear space Y. Then:

(1) The adjoint T^* of T is a continuous linear transformation of the Banach space Y^* into the Banach space X^*.
(2) If the range of T is dense in Y, then T^* is 1-1.
(3) Suppose X is a reflexive Banach space. If T is 1-1, then the range of T^* is dense in X^*.

PROOF. That T^* is linear is immediate. Further,

$$
\begin{aligned}
\|T^*(f)\| &= \sup_{\substack{x \in X \\ \|x\| \leq 1}} |[T^*(f)](x)| \\
&= \sup_{\substack{x \in X \\ \|x\| \leq 1}} |f(T(x))| \\
&\leq \sup_{\substack{x \in X \\ \|x\| \leq 1}} \|f\| \|T(x)\| \\
&\leq \|f\| \|T\|,
\end{aligned}
$$

showing that T^* is continuous in the norm topologies.

If $T^*(f) = 0$, then $f(T(x)) = 0$ for all $x \in X$. If the range of T is dense in Y, then $f(y) = 0$ for all $y \in Y$; i.e., f is the 0 functional, which implies that T^* is 1-1.

Now, if the range of T^* is not dense in X^*, then there exists a nonzero continuous linear functional ϕ on X^* such that ϕ is 0 on the range of T^*. (Why?) Therefore, $\phi(T^*(f)) = 0$ for all $f \in Y^*$. If X is reflexive, then $\phi = \hat{x}$ for some nonzero element $x \in X$. Therefore, $\hat{x}(T^*(f)) = [T^*(f)](x) = f(T(x)) = 0$ for every $f \in Y^*$. But then $T(x)$ belongs to the kernel of every element f in Y^*, whence $T(x)$ is the zero vector, which implies that T is not 1-1. Q.E.D.

THEOREM 5.6. Let X be a normed linear space, and let $\overline{B}_1$ denote the closed unit ball in the conjugate space X^* of X; i.e., $\overline{B}_1 = \{f \in X^* : \|f\| \leq 1\}$. Then:

(1) (Alaoglu) $\overline{B}_1$ is compact in the weak* topology on X^*.
(2) If X is separable, then $\overline{B}_1$ is metrizable in the weak* topology.

PROOF. By the definition of the weak* topology, we have that $\overline{B}_1$ is homeomorphic to a subset of the product space $\prod_{x \in X} \mathbb{R}$. See part e of Exercise 0.8. Indeed, the homeomorphism F is defined by

$$
[F(f)]_x = \hat{x}(f) = f(x).
$$

Since $|f(x)| \leq \|x\|$, for $f \in \overline{B}_1$, it follows in fact that

$$F(\overline{B}_1) \subseteq \prod_{x \in X} [-\|x\|, \|x\|],$$

which is a compact topological space K. Hence, to see that $\overline{B}_1$ is compact in the weak* topology, it will suffice to show that $F(\overline{B}_1)$ is a closed subset of K. Thus, if $\{f_\alpha\}$ is a net of elements of $\overline{B}_1$, for which the net $\{F(f_\alpha)\}$ converges in K to an element k, then $k_x = \lim[F(f_\alpha)]_x = \lim f_\alpha(x)$, for every x; i.e., the function f on X, defined by $f(x) = k_x$, is the pointwise limit of a net of linear functionals. Therefore f is itself a linear functional on X. Further, $|f(x)| \leq \|x\|$, implying that f is a continuous linear functional on X with $\|f\| \leq 1$, i.e., $f \in \overline{B}_1$. But then, the element $k \in K$ satisfies $k = F(f)$, showing that $F(\overline{B}_1)$ is closed in K, and this proves part 1.

Now, suppose that $\{x_n\}$ is a countable dense subset of X. Then

$$K^* = \prod_n [-\|x_n\|, \|x_n\|]$$

is a compact metric space, and the map $F^* : \overline{B}_1 \to K^*$, defined by

$$[F^*(f)]_n = f(x_n),$$

is continuous and 1-1, whence is a homeomorphism of $\overline{B}_1$ onto a compact metric space, and this completes the proof.

EXERCISE 5.15. (a) Prove that the closed unit ball in L^p is compact in the weak topology, for $1 < p < \infty$.

(b) Show that neither the closed unit ball in $L^1(\mathbb{R})$ nor the closed unit ball in c_0 is compact in its weak topology (or, in fact, in any locally convex vector space topology). HINT: Compact convex sets must have extreme points.

(c) Show that neither $L^1(\mathbb{R})$ nor c_0 is topologically isomorphic to the conjugate space of any normed linear space. Conclude that not every Banach space has a "predual."

(d) Conclude from part c that neither $L^1(\mathbb{R})$ nor c_0 is reflexive. Prove this assertion directly for $L^1(\mathbb{R})$ using part e of Exercise 5.2.

(e) Show that the closed unit ball in an infinite dimensional normed linear space is never compact in the norm topology.

THEOREM 5.7. (Critereon for a Banach Space to Be Reflexive) *Let X be a normed linear space, and let X^{**} denote its second dual equipped with the weak* topology. Then:*

(1) *$\hat{X}$, i.e., the set of all $\hat{x}$ for $x \in X$, is dense in $(X^{**}, \mathcal{W}^*)$.*

(2) *$\widehat{B_1}$, i.e., the set of all $\hat{x}$ for $\|x\| < 1$, is weak* dense in the closed unit ball V_1 of X^{**}.*

(3) *X is reflexive if and only if $\overline{B}_1$ is compact in the weak topology of X.*

PROOF. Suppose $\overline{\hat{X}}$ is a proper subspace of $(X^{**}, \mathcal{W}^*)$, and let ϕ be an element of X^{**} that is not in $\overline{\hat{X}}$. Since $\overline{\hat{X}}$ is a closed convex subspace in the weak* topology on X^{**}, there exists a weak* continuous linear functional η on $(X^{**}, \mathcal{W}^*)$ such that $\eta(\hat{x}) = 0$ for all $x \in X$ and $\eta(\phi) = 1$. By Theorem 5.2, every weak* continuous linear functional on X^{**} is given by an element of X^*. That is, there exists an $f \in X^*$ such that

$$\eta(\psi) = \hat{f}(\psi) = \psi(f)$$

for every $\psi \in X^{**}$. Hence,

$$f(x) = \hat{x}(f) = \eta(\hat{x}) = 0$$

for every $x \in X$, implying that $f = 0$. But,

$$\phi(f) = \hat{f}(\phi) = \eta(\phi) = 1,$$

implying that $f \neq 0$. Therefore, we have arrived at a contradiction, whence $\overline{\hat{X}} = X^{**}$ proving part 1.

We show part 2 in a similar fashion. Thus, suppose that $C = \overline{\widehat{B_1}}$ is a proper weak* closed (convex) subset of the norm closed unit ball V_1 of X^{**}, and let ϕ be an element of V_1 that is not an element of C. Again, since C is closed and convex, there exists by the Separation Theorem (Theorem 3.9) a weak* continuous linear functional η on $(X^{**}, \mathcal{W}^*)$ and a real number s such that $\eta(c) \leq s$ for all $c \in C$ and $\eta(\phi) > s$. Therefore, again by Theorem 5.2, there exists an $f \in X^*$ such that

$$\eta(\psi) = \psi(f)$$

for all $\psi \in X^{**}$. Hence,

$$f(x) = \hat{x}(f) = \eta(\hat{x}) \leq s$$

for all $x \in B_1$, implying that

$$|f(x)| \leq s$$

for all $x \in B_1$, and therefore that $\|f\| \leq s$. But, $\|\phi\| \leq 1$, and $\phi(f) = \eta(\phi) > s$, implying that $\|f\| > s$. Again, we have arrived at the desired contradiction, showing that $\widehat{B_1}$ is dense in V_1.

We have seen already that the map $x \to \hat{x}$ is continuous from $(X, \mathcal{W})$ into $(X^{**}, \mathcal{W}^*)$. See part d of Exercise 5.10. So, if $\overline{B}_1$ is weakly compact, then $\widehat{\overline{B}_1}$ is weak* compact in X^{**}, whence is closed in V_1. But, by part 2, $\widehat{\overline{B}_1}$ is dense in V_1, and so must equal V_1. It then follows immediately by scalar multiplication that $\hat{X} = X^{**}$, and X is reflexive.

Conversely, if X is reflexive, then the map $x \to \hat{x}$ is an isometric isomorphism, implying that $V_1 = \widehat{\overline{B}_1}$. Moreover, by Theorem 5.2, the map $x \to \hat{x}$ is a topological isomorphism of $(X, \mathcal{W})$ and $(X^{**}, \mathcal{W}^*)$. Since V_1 is weak* compact by Theorem 5.6, it then follows that $\overline{B}_1$ is weakly compact, and the proof is complete.

EXERCISE 5.16. Prove that every normed linear space is isometrically isomorphic to a subspace of some normed linear space $C(\Delta)$ of continuous functions on a compact Hausdorff space Δ. HINT: Use the map $x \to \hat{x}$.

We conclude this chapter by showing that Choquet's Theorem (Theorem 3.11) implies the Riesz Representation Theorem (Theorem 1.3) for compact metric spaces. Note, also, that we used the Riesz theorem in the proof of Choquet's theorem, so that these two results are really equivalent.

EXERCISE 5.17. (Choquet's Theorem and the Riesz Representation Theorem) Let Δ be a second countable compact topological space, and let $C(\Delta)$ denote the normed linear space of all continuous real-valued functions on Δ equipped with the supremum norm. Let K be the set of all continuous positive linear functionals ϕ on $C(\Delta)$ satisfying $\phi(1) = 1$.

(a) Show that K is compact in the weak* topology of $(C(\Delta))^*$.

(b) Show that the map $x \to \delta_x$ is a homeomorphism of Δ onto the set of extreme points of K. (Δ_x denotes the linear functional that sends f to the number $f(x)$.)

(c) Show that every positive linear functional on $C(\Delta)$ is continuous.

(d) Deduce the Riesz Representation Theorem in this case from Choquet's Theorem; i.e., show that every positive linear functional I on

$C(\Delta)$ is given by

$$I(f) = \int_{\Delta} f(x)\, d\mu(x),$$

where μ is a finite Borel measure on Δ.

CHAPTER VI

APPLICATIONS TO ANALYSIS

We include in this chapter several subjects from classical analysis to which the notions of functional analysis can be applied. Some of these subjects are essential to what follows in this text, e.g., convolution, approximate identities, and the Fourier transform. The remaining subjects of this chapter are highly recommended to the reader but will not specifically be referred to later.

Integral Operators

Let (S, μ) and (T, ν) be σ-finite measure spaces, and let k be a $\mu \times \nu$-measurable, complex-valued function on $S \times T$. We refer to the function k as a *kernel*, and we are frequently interested in when the formula

$$[K(f)](s) = \int_T k(s, t) f(t) \, d\nu(t) \qquad (6.1)$$

determines a bounded operator K from $L^p(\nu)$ into $L^r(\mu)$, for some $1 \le p \le \infty$ and some $1 \le r \le \infty$. Ordinarily, formula (6.1) is only valid for certain functions f, the so-called *domain* $D(K)$ of K. In any event, $D(K)$ is a vector space, and on this domain, K is clearly a linear transformation. More precisely, then, we are interested in when formula (6.1) determines a linear transformation K that can be extended to a bounded operator on all of $L^p(\nu)$ into $L^r(\mu)$. Usually, the domain $D(K)$ is a priori dense in $L^p(\nu)$, and the question above then reduces to whether K is a bounded operator from $D(K)$ into $L^r(\mu)$. That is, does there exist a constant M such that

$$\|K(f)\|_r = (\int_S |\int_T k(s,t)f(t)\,d\nu(t)|^r\,d\mu(s))^{1/r} \le M\|f\|_p$$

for all $f \in D(K)$. In such a case, we say that K is a *bounded integral operator*. In general, we say that the linear transformation K is an *integral operator determined by the kernel* $k(s,t)$.

The elementary result below is basically a consequence of Hoelder's inequality and the Fubini theorem.

THEOREM 6.1. *Suppose p, r are real numbers strictly between 1 and ∞, and let p' and r' satisfy*

$$1/p + 1/p' = 1/r + 1/r' = 1.$$

Suppose that $k(s,t)$ is a $\mu \times \nu$-measurable function on $S \times T$, and assume that the set $D(K)$ of all $f \in L^p(\nu)$ for which Equation (6.1) is defined is a dense subspace of $L^p(\nu)$. Then:

(1) *If the function b defined by $s \rightarrow \int_T |k(s,t)|^{p'}\,d\nu(t)$ is an element of $L^{r/p'}(\mu)$, then K is a bounded integral operator from $L^p(\nu)$ into $L^r(\mu)$.*

(2) *Suppose $p = r$ and that there exists an $\alpha \in [0,1]$ for which the function $s \rightarrow \int_T |k(s,t)|^{\alpha p'}\,d\nu(t)$ is an element of $L^\infty(\mu)$ with L^∞ norm c_1, and the function $t \rightarrow \int_S |k(s,t)|^{(1-\alpha)p}\,d\mu(s)$ is an element of $L^\infty(\nu)$ with L^∞ norm c_2. Then K is a bounded integral operator from $L^p(\nu)$ into $L^p(\mu)$.*

PROOF. Let $f \in D(K)$ be fixed. We have that

$$|\int_T k(s,t)f(t)\,d\nu(t)| \le \int_T |k(s,t)f(t)|\,d\nu(t)$$

$$\le (\int_T |k(s,t)|^{p'}\,d\nu(t))^{1/p'} \times (\int_T |f(u)|^p\,d\nu(u))^{1/p}$$

$$= b(s)^{1/p'}\|f\|_p,$$

from which it follows that

$$\|K(f)\|_r = (\int_S |\int_T k(s,t)f(t)\,d\nu(t)|^r\,d\mu(s))^{1/r} \le \|b\|_{r/p'}^{1/p'}\|f\|_p.$$

This proves part 1.

Again, for $f \in D(K)$ we have that

$$|\int_T k(s,t)f(t)\,d\nu(t)| \le \int_T |k(s,t)|^\alpha |k(s,t)|^{1-\alpha}|f(t)|\,d\nu(t)$$

$$\le (\int_T |k(s,t)|^{\alpha p'}\,d\nu(t))^{1/p'}$$

$$\times (\int_T |k(s,u)|^{(1-\alpha)p}|f(u)|^p\,d\nu(u))^{1/p}$$

$$\le c_1^{1/p'}(\int_T |k(s,u)|^{(1-\alpha)p}|f(u)|^p\,d\nu(u))^{1/p},$$

from which it follows that

$$\|K(f)\|_p = (\int_S |\int_T k(s,t)f(t)\,d\nu(t)|^p\,d\mu(s))^{1/p}$$

$$\le c_1^{1/p'}(\int_S \int_T |k(s,u)|^{(1-\alpha)p}|f(u)|^p\,d\nu(u)\,d\mu(s))^{1/p}$$

$$= c_1^{1/p'}(\int_T \int_S |k(s,u)|^{(1-\alpha)p}|f(u)|^p\,d\mu(s)\,d\nu(t))^{1/p}$$

$$\le c_1^{1/p'} c_2^{1/p}\|f\|_p,$$

and this proves part 2.

EXERCISE 6.1. (a) Restate part 1 of the above theorem for $p = r$.

(b) Restate part 1 of the above theorem for $r = p'$. Restate both parts of the theorem for $p = r = 2$.

(c) As a special case of part 2 of the theorem above, reprove it for $p = r = 2$ and $\alpha = 1/2$.

(d) How can we extend the theorem above to the case where p or r is 1 or ∞?

EXERCISE 6.2. Suppose both μ and ν are finite measures.

(a) Show that if the kernel $k(s,t)$ is a bounded function on $S \times T$, then (6.1) determines a bounded integral operator K for all p and r.

(b) Suppose $S = T = [a,b]$ and that μ and ν are both Lebesgue measure. Define k to be the characteristic function of the set of all pairs (s,t) for which $s \le t$. Show that (6.1) determines a bounded integral operator K from $L^1(\mu)$ into $L^1(\nu)$. Show further that $K(f)$ is always differentiable almost everywhere, and that $[K(f)]' = f$.

(c) Suppose k is an element of $L^2(\mu \times \nu)$. Use Theorem 6.1 to show that (6.1) determines a bounded integral operator K from $L^2(\nu)$ into $L^2(\mu)$.

(d) Is part c valid if μ and ν are only assumed to be σ-finite measures?

Convolution Kernels

THEOREM 6.2. (Young's Inequality) *Let f be a complex-valued measurable function on $\mathbb{R}^n$, and define $k \equiv k_f$ on $\mathbb{R}^n \times \mathbb{R}^n$ by*

$$k(x, y) = f(x - y).$$

If $f \in L^1(\mathbb{R}^n)$, then (6.1) determines a bounded integral operator $K \equiv K_f$ from $L^p(\mathbb{R}^n)$ into itself, where we equip each space $\mathbb{R}^n$ with Lebesgue measure. Moreover, $\|K_f(g)\|_p \leq \|f\|_1 \|g\|_p$ for every $g \in L^p(\mathbb{R}^n)$.

PROOF. Assume first that f is continuous with compact support. Let g be in L^p and h be in $L^{p'}$, $(1/p + 1/p' = 1)$. By Tonelli's Theorem, we have that

$$
\begin{aligned}
\int_{\mathbb{R}^n} \int_{\mathbb{R}^n} |f(x-y)g(y)h(x)|\, dy dx &= \int_{\mathbb{R}^n} \int_{\mathbb{R}^n} |f(-y)g(x+y)h(x)|\, dy dx \\
&\leq \|f\|_1 \sup_y \int_{\mathbb{R}^n} |g(x+y)h(x)|\, dx \\
&\leq \|f\|_1 \sup_y \|g\|_p \|h\|_{p'},
\end{aligned}
$$

which shows that the function $f(x-y)g(y)h(x)$ is integrable on $\mathbb{R}^n \times \mathbb{R}^n$. Therefore, for almost all x, the function $f(x-y)g(y)$ is integrable on $\mathbb{R}^n$, and the resulting function $K_f(g)$ of x belongs to L^p. Moreover,

$$\|K_f(g)\|_p \leq \|f\|_1 \|g\|_p.$$

Now let f be an arbitrary element of L^1, and let $\{f_n\}$ be a sequence of continuous functions having compact supports such that $f = \lim f_n$ in L^1. For each $g \in L^p$, the sequence $\{K_{f_n}(g)\}$ is a Cauchy sequence in L^p by the preceding paragraph. Writing $g' = \lim K_{f_n}(g)$, and letting h be an arbitrary element of $L^{p'}$, we have that

$$
\begin{aligned}
\int_{\mathbb{R}^n} g'(x)h(x)\, dx &= \lim \int_{\mathbb{R}^n} \int_{\mathbb{R}^n} f_n(x-y)g(y)h(x)\, dy dx \\
&= \lim \int_{\mathbb{R}^n} \int_{\mathbb{R}^n} f_n(-y)g(x+y)h(x)\, dy dx \\
&= \lim \int_{\mathbb{R}^n} \int_{\mathbb{R}^n} f_n(-y)g(x+y)h(x)\, dx dy \\
&= \int_{\mathbb{R}^n} \int_{\mathbb{R}^n} f(-y)g(x+y)h(x)\, dx dy,
\end{aligned}
$$

where the last equality follows from the two facts: $f = \lim f_n$ in L^1, and the function $y \to \int g(x + y)h(x)\,dx$ is uniformly bounded. Hence, for almost all x, the function $f(-y)g(x + y)$ is integrable and

$$g'(x) = \int_{\mathbb{R}^n} f(-y)g(x + y)\,dy = \int_{\mathbb{R}^n} f(x - y)g(y)\,dy$$

for almost all x. So, K_f is a linear transformation of L^p into L^p. Moreover, the above calculations imply that

$$\|K_f(g)\|_p \leq \lim \|f_n\|_1 \|g\|_p = \|f\|_1 \|g\|_p,$$

showing that K_f is a bounded integral operator on L^p and establishes the desired inequality.

By $\mathbb{T}$ we shall mean the half-open interval $[0,1)$, and we shall refer to $\mathbb{T}$ as the *circle*. By $L^p(\mathbb{T})$ we shall mean the set of all Lebesgue measurable functions on $\mathbb{R}$, which are periodic with period 1, satisfying $\int_0^1 |f(x)|^p\,dx < \infty$.

EXERCISE 6.3 (a) Use part 2 of Theorem 6.1 to give an alternative proof of Theorem 6.2.

(b) (Convolution on the circle) If $f \in L^1(\mathbb{T})$, define $k = k_f$ on $\mathbb{T} \times \mathbb{T}$ by $k_f(x,y) = f(x - y)$. Prove that K_f is a bounded integral operator from $L^p(\mathbb{T})$ into itself for all $1 \leq p \leq \infty$. In fact, prove this two ways: Use Theorem 6.1, and then mimic the proof of Theorem 6.2.

DEFINITION. If $f \in L^1(\mathbb{R}^n)$ $(L^1(\mathbb{T}))$, then the bounded integral operator K_f of the preceding theorem (exercise) is called the *convolution operator* by f, and we denote $K_f(g)$ by $f * g$. The kernel $k_f(x,y) = f(x - y)$ is called a *convolution kernel*.

EXERCISE 6.4. (a) Suppose $f \in L^p(\mathbb{R}^n)$ and $g \in L^{p'}(\mathbb{R}^n)$. Show that the function $f * g$, defined by

$$(f * g)(x) = \int_{\mathbb{R}^n} f(x - y)g(y)\,dy,$$

is everywhere well-defined. Show further that $f * g$ is continuous and vanishes at infinity. Show finally that $f * g = g * f$.

(b) If $f, g, h \in L^1$, show that $f * g = g * f$ and that $(f * g) * h = f * (g * h)$.

The next result is a useful generalization of Theorem 6.2.

THEOREM 6.3. *Let f be an element of $L^p(\mathbb{R}^n)$. Then, for any $1 \le q \le p'$, convolution by f is a bounded operator from $L^q(\mathbb{R}^n)$ into $L^r(\mathbb{R}^n)$, where $1/p + 1/q - 1/r = 1$.*

EXERCISE 6.5. Use the Riesz Interpolation Theorem, Theorem 6.2, and Exercise 6.4 to prove Theorem 6.3.

REMARK. Later, we will be interested in convolution kernels k_f where the function f does not belong to any L^p space. Such kernels are called *singular kernels*. Though the arguments above cannot be used on such singular kernels, nevertheless these kernels often define bounded integral operators.

Reproducing Kernels and Approximate Identities

DEFINITION. Let (S, μ) be a σ-finite measure space and let $k(x, y)$ be a $\mu \times \mu$-measurable kernel on $S \times S$. Suppose that the operator K, defined by (6.1), is a bounded integral operator from $L^p(\mu)$ into itself. Then K is called a *reproducing kernel* for a subspace V of $L^p(\mu)$ if $K(g) = g$ for all $g \in V$. A parameterized family $\{k_t\}$ of kernels is called an *approximate identity* for a subspace V of $L^p(\mu)$ if all the corresponding operators K_t are bounded integral operators, and $\lim_{t \to 0} K_t(g) = g$ for every $g \in V$, where the limit is taken in $L^p(\mu)$.

THEOREM 6.4. *Let S be the closed unit disk in $\mathbb{C}$, let μ be the measure on S defined on the space $C(S)$ of continuous functions on S by*

$$\int_S f(z)\,d\mu(z) = \int_0^{2\pi} f(e^{i\theta})\,d\theta.$$

Let $p = 1$, and let H be the subspace of $L^1(\mu)$ consisting of the (complex-valued) functions that are continuous on S and analytic on the interior of S. Let $k(z, \zeta)$ be the kernel on $S \times S$ defined by

$$k(z, \zeta) = \frac{1}{2\pi} \frac{1}{1 - (z/\zeta)},$$

if $z \ne \zeta$, and

$$k(z, z) = 0$$

for all $z \in S$. Then k is a reproducing kernel for H.

EXERCISE 6.6. Prove Theorem 6.4. HINT: Cauchy's formula.

REMARK. Among the most interesting reproducing kernels and approximate identities are the ones that are convolution kernels.

THEOREM 6.5. *Let k be a nonnegative Lebesgue-measurable function on $\mathbb{R}^n$ for which $\int k(x)\, dx = 1$. For each positive t, define*

$$k_t(x) = (1/t^n)k(x/t),$$

and set

$$K(x) = \int_{\|x\| \leq \|y\|} k(y)\, dy.$$

Then:
  (1) *If f is uniformly continuous and bounded on $\mathbb{R}^n$, then $k_t * f$ converges uniformly to f on $\mathbb{R}^n$ as t approaches 0.*
  (2) *If $K \in L^p(\mathbb{R}^n)$ $(1 \leq p < \infty)$, then $k_t * f$ converges to f in $L^p(\mathbb{R}^n)$ for every $f \in L^p(\mathbb{R}^n)$ as t approaches 0.*
  (3) *If $k \in L^{p'}(\mathbb{R}^n)$, $f \in L^p(\mathbb{R}^n)$ $(1 \leq p < \infty$ and $1/p + 1/p' = 1)$, and f is continuous at a point x, then $(k_t * f)(x)$ converges to $f(x)$ as t approaches 0.*

PROOF. To prove part 1, we must show that for each $\epsilon > 0$ there exists a $\delta > 0$ such that if $t < \delta$ then $|(k_t * f)(x) - f(x)| < \epsilon$ for all x. Note first that $\int k_t(x)\, dx = 1$ for all t. Write

$$(k_t * f)(x) = \int k_t(x - y)f(y)\, dy = \int k_t(y)f(x - y)\, dy.$$

So, we have that

$$|(k_t * f)(x) - f(x)| = \left| \int k_t(y)f(x - y)\, dy - f(x)\int k_t(y)\, dy \right|$$

$$\leq \int k_t(y)|f(x - y) - f(x)|\, dy$$

$$= \int_{\|y\| \leq h} k_t(y)|f(x - y) - f(x)|\, dy$$

$$+ \int_{\|y\| > h} k_t(y)|f(x - y) - f(x)|\, dy$$

for any positive h. Therefore, given $\epsilon > 0$, choose h so that $|f(w) - f(z)| < \epsilon/2$ if $\|w - z\| < h$, and set $M = \|f\|_\infty$. Then

$$|(k_t * f)(x) - f(x)| \leq \int_{\|y\| \leq h} k_t(y)(\epsilon/2)\, dy + \int_{\|y\| > h} k_t(y)2M\, dy$$

$$\leq (\epsilon/2) \int k_t(y)\, dy + 2M \int_{\|y\| > h} k_t(y)\, dy$$

$$= (\epsilon/2) + 2M \int_{\|y\| > h/t} k(y)\, dy$$

for all x. Finally, since $k \in L^1(\mathbb{R}^n)$, there exists a $\rho > 0$ such that

$$\int_{\|y\| > \rho} k(y) \, dy < \epsilon/(4M),$$

whence

$$|(k_t * f)(x) - f(x)| < \epsilon$$

for all x if $t < \delta = h/\rho$. This proves part 1.

By Theorem 6.2 we have that $\|k_t * f\|_p \le \|f\|_p$ for all t. Hence, if $f \in L^p(\mathbb{R}^n)$ and $\{f_j\}$ is a sequence of continuous functions with compact support that converges to f in L^p norm, then

$$\|k_t * f - f\|_p \le \|k_t * (f - f_j)\|_p + \|k_t * f_j - f_j\|_p + \|f_j - f\|_p.$$

Given $\epsilon > 0$, choose j so that the first and third terms are each bounded by $\epsilon/3$. Hence, we need only verify part 2 for an $f \in L^p(\mathbb{R}^n)$ that is continuous and has compact support. Suppose the support of such an f is contained in the ball of radius a around 0. From the proof above for part 1, we see that $|(k_t * f)(x) - f(x)| \le 2M$ for all x. Moreover, if $\|x\| \ge 2a$ and $t < 1/2$, then

$$|(k_t * f)(x) - f(x)| = |\int k_t(y)(f(x - y) - f(x)) \, dy|$$

$$\le \int k_t(y)|f(x - y)| \, dy$$

$$= \int_{\|x\| - a \le \|y\| \le \|x\| + a} k_t(y)|f(x - y)| \, dy$$

$$\le M \int_{\|x\| - a \le \|y\|} k_t(y) \, dy$$

$$\le M \int_{\|x/2\| \le \|y\|} k_t(y) \, dy$$

$$= M \int_{\|x/2t\| \le \|y\|} k(y) \, dy$$

$$\le M \int_{\|x\| \le \|y\|} k(y) \, dy$$

$$= M K(x).$$

Hence, $|(k_t * f)(x) - f(x)|$ is bounded for all t by a fixed function in $L^p(\mathbb{R}^n)$, so that part 2 follows from part 1 and the dominated convergence theorem.

We leave part 3 to the exercises.

EXERCISE 6.7. (a) Prove part 3 of the preceding theorem.

(b) (Poisson Kernel on the Line) For each $t > 0$ define a kernel k_t on $\mathbb{R} \times \mathbb{R}$ by

$$k_t(x, y) = (t/\pi)(1/(t^2 + (x - y)^2)).$$

Prove that $\{k_t\}$ is an approximate identity for $L^p(\mathbb{R})$ for $1 \le p < \infty$. HINT: Note that the theorem above does not apply directly. Alter the proof.

(c) (Poisson Kernel in $\mathbb{R}^n$) Let $c = \int_{\mathbb{R}^n}(1/(1 + \|x\|^2)^{(n+1)/2})\, dx$. For each positive t, define a kernel k_t on $\mathbb{R}^n \times \mathbb{R}^n$ by

$$k_t(x, y) = \frac{t/c}{(t^2 + \|x - y\|^2)^{(n+1)/2}}.$$

Prove that $\{k_t\}$ is an approximate identity for $L^p(\mathbb{R}^n)$ for $1 \le p < \infty$.

(d) (Poisson Kernel on the Circle) For each $0 < r < 1$ define a function k_r on $\mathbb{T}$ by

$$k_r(x) = \frac{1 - r^2}{1 + r^2 - 2r\cos(2\pi x)}.$$

Show that $k_r(x) \ge 0$ for all r and x, and that

$$k_r(x) = \sum_{n=-\infty}^{\infty} r^{|n|}e^{2\pi inx},$$

whence $\int_0^1 k_r(x)\, dx = 1$ for every $0 < r < 1$. Prove that $\{k_r\}$ is an approximate identity for $L^p(\mathbb{T})$ ($1 \le p < \infty$) in the sense that

$$f = \lim_{r \to 1} k_r * f,$$

where the limit is taken in $L^p(\mathbb{T})$.

EXERCISE 6.8. (Gauss Kernel) (a) Define g on $\mathbb{R}$ by

$$g(x) = (1/\sqrt{2\pi})e^{-x^2/2},$$

and set

$$g_t(x) = (1/\sqrt{t})g(x/\sqrt{t}) = (1/\sqrt{2\pi t})e^{-x^2/2t}.$$

Prove that $\{g_t\}$ is an approximate identity for $L^p(\mathbb{R})$ for $1 \le p < \infty$.

(b) Define g on $\mathbb{R}^n$ by

$$g(x) = (1/(2\pi)^{n/2})e^{-\|x\|^2/2},$$

and set

$$g_t(x) = (1/t^{n/2})g(x/\sqrt{t}) = (2\pi t)^{-n/2}e^{-\|x\|^2/2t}.$$

Prove that $\{g_t\}$ is an approximate identity for $L^p(\mathbb{R}^n)$ for $1 \le p < \infty$.

Green's Functions

DEFINITION. Let μ be a σ-finite Borel measure on $\mathbb{R}^n$, let D be a dense subspace of $L^p(\mu)$, and suppose L is a (not necessarily continuous) linear transformation of D into $L^p(\mu)$. By a *Green's function for L* we shall mean a $\mu \times \mu$-measurable kernel $g(x, y)$ on $\mathbb{R}^n \times \mathbb{R}^n$, for which the corresponding (not necessarily bounded) integral operator G satisfies the following: If v belongs to the range of L, then $G(v)$, defined by

$$[G(v)](x) = \int_{\mathbb{R}^n} g(x, y)v(y)\, d\mu(y),$$

belongs to D, and $L(G(v)) = v$. That is, the integral operator G is a right inverse for the transformation L.

Obviously, knowing a Green's function for an operator L is of use in solving for u in an equation like $L(u) = f$. Not every (even invertible) linear transformation L has a Green's function, although many classical transformations do. There are various techniques for determining Green's functions for general kinds of transformations L, but the most important L's are differential operators. The following exercise gives a classical example of the construction of a Green's function for such a transformation.

EXERCISE 6.9. Let b be a positive real number, let f be an element of $L^1([0, b])$, and consider the nth order ordinary differential equation:

$$u^{(n)} + a_{n-1}u^{(n-1)} + \ldots + a_1 u' + a_0 u = f, \qquad (6.2)$$

where the coefficients $a_0, \ldots, a_{n-1}$ are constants. Let D denote the set of all n times everywhere-differentiable functions u on $[0, b]$ for which $u^{(n)} \in L^1([0, b])$, and let L be the transformation of D into $L^1([0, b]) \subset L^1(\mathbb{R})$ defined by

$$L(u) = u^{(n)} + a_{n-1}u^{(n-1)} + \ldots + a_1 u' + a_0 u.$$

Let A denote the $n \times n$ matrix defined by

$$A = \begin{bmatrix} 0 & 1 & 0 & \ldots & 0 \\ 0 & 0 & 1 & \ldots & 0 \\ \ldots & \ldots & \ldots & \ldots & \ldots \\ -a_0 & -a_1 & -a_2 & \ldots & -a_{n-1} \end{bmatrix},$$

let $\vec{F}(t)$ be the vector-valued function given by

$$\vec{F}(t) = \begin{bmatrix} 0 \\ \cdot \\ \cdot \\ \cdot \\ 0 \\ f(t) \end{bmatrix},$$

and consider the vector-valued differential equation

$$\dot{\vec{U}} = A\vec{U} + \vec{F}. \tag{6.3}$$

(a) Show that if $\vec{U}$ is a solution of Equation (6.3), then u_1 is a solution of Equation (6.2), where u_1 is the first component of $\vec{U}$.

(b) If B is an $n \times n$ matrix, write e^B for the matrix defined by

$$e^B = \sum_{j=0}^{\infty} B^j/j!.$$

Define $\vec{U}$ on $[0, b]$ by

$$\vec{U}(t) = \int_0^t e^{(t-s)A} \times \vec{F}(s)\, ds.$$

Prove that $\vec{U}$ is a solution of Equation (6.3).

(c) For $1 \leq i, j \leq n$, write $c_{ij}(t)$ for the ijth component of the matrix e^{tA}. Define $g(t, s) = c_{1n}(t - s)$ if $s \leq t$ and $g(t, s) = 0$ otherwise. Prove that g is a Green's function for L.

We give next two general, but certainly not all-inclusive, results on the existence of Green's functions.

If $h(x, y)$ is a function of two variables, we denote by h^y the function of x defined by $h^y(x) = h(x, y)$.

THEOREM 6.6. Let μ be a regular (finite on compact sets) σ-finite Borel measure on $\mathbb{R}^n$, let D be a dense subspace of $L^p(\mu)$ $(1 \le p \le \infty)$, and let L be a (not necessarily continuous) linear transformation of D into $L^1(\mu)$. Assume that:

(1) There exists a bounded integral operator K from $L^1(\mu)$ into $L^1(\mu)$, determined by a kernel $k(x,y)$, for which k is a reproducing kernel for the range V of L, and such that the map $y \to k^y$ is uniformly continuous from $\mathbb{R}^n$ into $L^1(\mu)$.

(2) There exists a bounded integral operator G from $L^1(\mu)$ into $L^p(\mu)$, determined by a kernel $g(x,y)$, such that the map $y \to g^y$ is uniformly continuous from $\mathbb{R}^n$ into D, and such that $L(g^y) = k^y$ for all y.

(3) The graph of L, thought of as a subset of $L^p(\mu) \times L^1(\mu)$, is closed. That is, if $\{u_j\}$ is a sequence of elements of D that converges to an element $u \in L^p$, and if the sequence $\{L(u_j)\}$ converges in L^1 to a function v, then the pair (u,v) belongs to the graph of L; i.e., $u \in D$ and $v = L(u)$.

Then g is a Green's function for L.

PROOF. Let v be in the range of L, and let $\{\phi_j\}$ be a sequence of simple functions having compact support that converges to v in $L^1(\mu)$. Because μ is regular and σ-finite, we may assume that

$$\phi_j = \sum_{i=1}^{n_j} a_{i,j}\chi_{E_{i,j}},$$

where

$$\lim_{j\to\infty}\max_i \operatorname{diam}(E_{i,j}) \equiv \lim \delta_j = 0.$$

For each $j = 1,2,\ldots$ and each $1 \le i \le n_j$, let $y_{i,j}$ be an element of $E_{i,j}$, and define functions v_j and u_j by

$$v_j(x) = \sum_{i=1}^{n_j} a_{i,j}\mu(E_{i,j})k^{y_{i,j}}(x) = \sum_{i=1}^{n_j} a_{i,j}\mu(E_{i,j})k(x,y_{i,j})$$

and

$$u_j(x) = \sum_{i=1}^{n_j} a_{i,j}\mu(E_{i,j})g^{y_{i,j}}(x). = \sum_{i=1}^{n_j} a_{i,j}\mu(E_{i,j})g(x,y_{i,j}).$$

Notice that each $u_j \in D$ and that $v_j = L(u_j)$. Finally, for each positive δ, define $\epsilon_1(\delta)$ and $\epsilon_2(\delta)$ by

$$\epsilon_1(\delta) = \sup_{\|y-y'\|<\delta} \|k^y - k^{y'}\|_1$$

and

$$\epsilon_2(\delta) = \sup_{\|y-y'\|<\delta} \|g^y - g^{y'}\|_p.$$

By the uniform continuity assumptions on the maps $y \to k^y$ and $y \to g^y$, we know that

$$0 = \lim_{\delta \to 0} \epsilon_i(\delta).$$

First, we have that $v = \lim_j v_j$. For

$$\|v - v_j\|_1 = \|K(v) - v_j\|_1$$
$$\leq \|K(v - \phi_j)\|_1 + \|K(\phi_j) - v_j\|_1$$
$$\leq \|K\|\|v - \phi_j\|_1 + \int |[K(\phi_j)](x) - v_j(x)|\, d\mu(x)$$
$$= \|K\|\|v - \phi_j\|_1 + \int |\int k(x,y) \sum_{i=1}^{n_j} a_{i,j}\chi_{E_{i,j}}(y)\, dy$$
$$- \sum_{i=1}^{n_j} a_{i,j}\mu(E_{i,j})k(x,y_{i,j})|\, dx$$
$$= \|K\|\|v - \phi_j\|_1 + \int |\sum_{i=1}^{n_j}[\int k(x,y)a_{i,j}\chi_{E_{i,j}}(y)$$
$$- k(x,y_{i,j})a_{i,j}\chi_{E_{i,j}}(y)\, dy]|\, dx$$
$$\leq \|K\|\|v - \phi_j\|_1$$
$$+ \sum_{i=1}^{n_j} |a_{i,j}| \int \chi_{E_{i,j}}(y) \int |k(x,y) - k(x,y_{i,j})|\, dxdy$$
$$= \|K\|\|v - \phi_j\|_1 + \sum_{i=1}^{n_j} |a_{i,j}| \int \|k^y - k^{y_{i,j}}\|_1 \chi_{E_{i,j}}(y)\, dy$$
$$\leq \|K\|\|v - \phi_j\|_1 + \epsilon_1(\delta_j)\|\phi_j\|_1,$$

which tends to zero as j tends to ∞.

Similarly, we have that $G(v) = \lim_j u_j$. (See the following exercise.) So, since the graph of L is closed, and since $L(u_j) = v_j$ for all j, we see that $G(v) \in D$ and $L(G(v)) = v$, as desired.

EXERCISE 6.10. In the proof of Theorem 6.6, verify that $G(v)$ is the L^p limit of the sequence $\{u_j\}$.

THEOREM 6.7. *Let μ, D, and L be as in the preceding theorem. Suppose $\{g_t(x, y)\}$ is a parameterized family of kernels on $\mathbb{R}^n \times \mathbb{R}^n$ such that for each t the map $y \to g_t^y$ is uniformly continuous from $\mathbb{R}^n$ into D. Suppose that $\{k_t(x, y)\}$ is an approximate identity for the range of L, that for each t the map $y \to k_t^y$ is uniformly continuous from $\mathbb{R}^n$ into $L^1(\mu)$, and that $L(g_t^y) = k_t^y$ for all t and y. Suppose finally that $\lim_{t \to 0} g_t(x, y) = g(x, y)$ for almost all x and y, and that $\lim_{t \to 0} G_t(v) = G(v)$ for each v in the range of L, where G_t and G are the integral operators determined by the kernels g_t and g respectively. Then g is a Green's function for L.*

EXERCISE 6.11. Prove Theorem 6.7. HINT: For v in the range of L, show that $G_t(v) \in D$ and that $L(G_t(v)) = K_t(v)$. Then use again the fact that the graph of L is closed.

EXERCISE 6.12. Let μ be Lebesgue measure on $\mathbb{R}^n$, and suppose D and L are as in the preceding two theorems. Assume that L is homogeneous of degree d. That is, if δ_t is the map of $\mathbb{R}^n$ into itself defined by $\delta_t(x) = tx$, then

$$L(u \circ \delta_t) = t^d[L(u)] \circ \delta_t.$$

(Homogeneous differential operators fall into this class.) Suppose p is a nonnegative function on $\mathbb{R}^n$ of integral 1, and that u_0 is an element of D for which $L(u_0) = p$. Define $g_t(x) = t^{d-n} u_0(x/t)$, and assume that, for each v in the range of L, $\lim_{t \to 0} g_t * v$ exists and that g_t converges, as t approaches 0, almost everywhere to a function g. Show that g is a Green's function for L. HINT: Use Theorem 6.5 to construct an approximate identity from the function p. Then verify that the hypotheses of Theorem 6.7 hold.

Fourier Transform

DEFINITION. If f is a complex-valued function in $L^1(\mathbb{R})$, define a function $\hat{f}$ on $\mathbb{R}$ by

$$\hat{f}(\xi) = \int_{-\infty}^{\infty} f(x) e^{-2\pi i x \xi} \, dx.$$

The function $\hat{f}$ is called the *Fourier transform* of f.

EXERCISE 6.13. (a) (Riemann-Lebesgue Theorem) For $f \in L^1$, show that the Fourier transform $\hat{f}$ of f is continuous, vanishes at infinity, and $\|\hat{f}\|_\infty \leq \|f\|_1$. HINT: Do this first for f the characteristic function of a finite interval (a, b) and then approximate (in L^1 norm) an arbitrary f by step functions.

(b) (Convolution Theorem) If f and g are elements of L^1, show that

$$\widehat{f * g} = \hat{f}\hat{g}$$

and

$$\int f\hat{g} = \int \hat{f}g.$$

(c) For $f \in L^1$, define f^* by

$$f^*(x) = \overline{f(-x)}.$$

Show that $\widehat{f^*} = \overline{\hat{f}}$.

(d) If $|x||f(x)| \in L^1$, show that $\hat{f}$ is differentiable, and

$$\hat{f}'(\xi) = -2\pi i \int x f(x) e^{-2\pi i x \xi} \, dx.$$

(e) If f is absolutely continuous $(f(x) = \int_{-\infty}^{x} f')$, and both f and f' are in $L^1(\mathbb{R})$, show that $\xi \hat{f}(\xi) \in C_0$.

(f) Show that the Fourier transform sends Schwartz space $\mathcal{S}$ into itself.

(g) If $f(x) = e^{-2\pi|x|}$, show that

$$\hat{f}(\xi) = (1/\pi)\frac{1}{1 + \xi^2}.$$

(h) If $g(x) = e^{-\pi x^2}$, show that

$$\hat{g}(\xi) = e^{-\pi \xi^2} = g(\xi).$$

That is, $\hat{g} = g$. HINT: Show that $\hat{g}$ satisfies the differential equation

$$\hat{g}'(\xi) = -2\pi \xi \hat{g}(\xi),$$

and

$$\hat{g}(0) = 1.$$

Recall that

$$\int_{-\infty}^{\infty} e^{-x^2/2}\,dx = \sqrt{2\pi}.$$

EXERCISE 6.14. (Inversion Theorem)

(a) (Fourier transform of the Gauss kernel) If g_t is the function defined by

$$g_t(x) = (1/\sqrt{2\pi t})e^{-x^2/2t},$$

use part h of the preceding exercise to show that

$$\widehat{g}_t(\xi) = e^{-2\pi t\xi^2},$$

whence

$$g_t(x) = \int \widehat{g}_t(\xi)e^{2\pi i x\xi}\,d\xi.$$

(b) Show that for any $f \in L^1$, for which $\widehat{f}$ also is in L^1, we have that f is continuous and

$$f(x) = \int \widehat{f}(\xi)e^{2\pi i x\xi}\,d\xi.$$

HINT: Make use of the fact that the g_t's of part a form an approximate identity. Establish the equality

$$\int \widehat{f}(\xi)e^{2\pi i x\xi}\,d\xi = \lim_{t\to 0} \int \widehat{f}(\xi)\widehat{g}_t(\xi)e^{2\pi i x\xi}\,d\xi,$$

and then use the convolution theorem.

(c) Conclude that the Fourier transform is 1-1 on L^1.

(d) Show that Schwartz space is mapped 1-1 and onto itself by the Fourier transform. Show further that the Fourier transform is a topological isomorphism of order 4 from the locally convex topological vector space $\mathcal{S}$ onto itself.

THEOREM 6.8. (Plancherel Theorem) If $f \in L^1(\mathbb{R})\cap L^2(\mathbb{R})$, then $\widehat{f} \in L^2(\mathbb{R})$ and $\|f\|_2 = \|\widehat{f}\|_2$. Consequently, if $f,g \in L^1(\mathbb{R})\cap L^2(\mathbb{R})$, then

$$\int f(x)\overline{g(x)}\,dx = \int \widehat{f}(\xi)\overline{\widehat{g}(\xi)}\,d\xi.$$

PROOF. Suppose first that f is in Schwartz space $\mathcal{S}$, and write f^* for the function defined by

$$f^*(x) = \overline{f(-x)}.$$

Then $f * f^* \in L^1$, and $\widehat{f * f^*} = |\hat{f}|^2 \in L^1$. So, by the Inversion Theorem,

$$\begin{aligned}
\|f\|_2^2 &= \int f(x)\overline{f(x)}\, dx \\
&= \int f(x)f^*(-x)\, dx \\
&= (f * f^*)(0) \\
&= \int \widehat{f * f^*}(\xi) e^{2\pi i 0 \times \xi}\, d\xi \\
&= \int \widehat{f * f^*}(\xi)\, d\xi \\
&= \int |\hat{f}(\xi)|^2\, d\xi \\
&= \|\hat{f}\|_2^2.
\end{aligned}$$

Now, if f is an arbitrary element of $L^1(\mathbb{R}) \cap L^2(\mathbb{R})$, and if $\{f_n\}$ is a sequence of elements of $\mathcal{S}$ that converges to f in L^2 norm, then the sequence $\{\widehat{f_n}\}$ is a Cauchy sequence in $L^2(\mathbb{R})$, whence converges to an element $g \in L^2(\mathbb{R})$. We need only show that g and $\hat{f}$ agree almost everywhere. If h is any element of $L^1(\mathbb{R}) \cap L^2(\mathbb{R})$ we have, using part b of Exercise 6.13, that

$$\begin{aligned}
\int g(\xi)h(\xi)\, d\xi &= \lim \int \widehat{f_n}(\xi)h(\xi)\, d\xi \\
&= \lim \int f_n(\xi)\hat{h}(\xi)\, d\xi \\
&= \int f(\xi)\hat{h}(\xi)\, d\xi \\
&= \int \hat{f}(\xi)h(\xi)\, d\xi,
\end{aligned}$$

showing that $\hat{f}$ and g agree as L^2 functions. (Why?) It follows then that $\hat{f} \in L^2$ and $\|\hat{f}\|_2 = \|f\|_2$.

The final equality of the theorem now follows from the polarization identity in $L^2(\mathbb{R})$. That is, for any $f, g \in L^2(\mathbb{R})$, we have

$$\int f\bar{g} = (1/4)\sum_{k=0}^{3} i^k \int (f + i^k g)\overline{(f + i^k g)},$$

which can be verified by expanding the right-hand side.

REMARK. The Plancherel theorem asserts that the Fourier transform is an isometry in the L^1 norm from $L^1(\mathbb{R})\cap L^2(\mathbb{R})$ into $L^2(\mathbb{R})$. Since Schwartz space is in the range of the Fourier transform on $L^1(\mathbb{R})\cap L^2(\mathbb{R})$, the Fourier transform maps $L^1(\mathbb{R}) \cap L^2(\mathbb{R})$ onto a dense subspace of $L^2(\mathbb{R})$, whence there exists a unique extension U of the Fourier transform from $L^1(\mathbb{R})\cap L^2(\mathbb{R})$ to an isometry on all of $L^2(\mathbb{R})$. This U is called the L^2 *Fourier transform.* It is an isometry of $L^2(\mathbb{R})$ onto itself.

EXERCISE 6.15. (a) Suppose $f(x)$ and $xf(x)$ are both elements of $L^2(\mathbb{R})$. Prove that $U(f)$ is differentiable and compute $[U(f)]'(\xi)$.

(b) If f is absolutely continuous and both f and f' belong to $L^2(\mathbb{R})$, show that $f(x) = \int_{-\infty}^{x} f'(t)\,dt$ and that $[U(f')](\xi) = 2\pi i\xi[U(f)](\xi)$. State and prove results for the L^2 Fourier transform that are analogous to parts b and c of Exercise 6.13.

(c) Suppose f is in $L^2(\mathbb{R})$ but not $L^1(\mathbb{R})$. Assume that for almost every ξ, the function $f(x)e^{-2\pi i x\xi}$ is improperly Riemann integrable. That is, assume that there exists a function g such that

$$\lim_{B\to\infty} \int_{-B}^{B} f(x)e^{-2\pi i x\xi}\,dx$$

exists and equals $g(\xi)$ for almost all ξ. Prove that $g = U(f)$.

(d) Define the function f by $f(x) = \sin(x)/x$. Prove that $f \in L^2(\mathbb{R})$ but not in $L^1(\mathbb{R})$. Show that f is improperly Riemann integrable, and establish that

$$\lim_{B\to\infty} \int_{-B}^{B} f(x)\,dx = \pi.$$

HINT: Verify the following equalities:

$$\lim_{B \to \infty} \int_{-B}^{B} f(x)\,dx = 2\lim_{n} \int_{0}^{\pi(n+1/2)} f(x)\,dx$$

$$= \lim_{n} \int_{0}^{\pi} \frac{\sin((n+1/2)x)}{x/2}\,dx$$

$$= \lim_{n} \int_{0}^{\pi} \frac{\sin((n+1/2)x)}{\sin((1/2)x)}\,dx$$

$$= \lim_{n} \int_{0}^{\pi} \frac{e^{-i(n+1/2)x} - e^{i(n+1/2)x}}{e^{-i(1/2)x} - e^{i(1/2)x}}\,dx$$

$$= \lim_{n} \int_{0}^{\pi} \sum_{k=-n}^{n} e^{ikx}\,dx$$

$$= \pi.$$

(e) Fix a $\delta > 0$, and let $f_\delta(x) = 1/x$ for $|x| \geq \delta$. Use part c to show that

$$[U(f_\delta)](\xi) = -i\mathrm{sgn}(\xi) \lim_{B \to \infty} \int_{2\pi\delta|\xi| \leq |x| \leq B} \sin(x)/x\,dx,$$

where sgn denotes the *signum* function defined on $\mathbb{R}$ by

$$\mathrm{sgn}(t) = 1, \text{ for } t > 0$$

$$\mathrm{sgn}(0) = 0$$

$$\mathrm{sgn}(t) = -1, \text{ for } t < 0.$$

Using part d, conclude that $[U(f_\delta)](\xi)$ is uniformly bounded in both the variables δ and ξ, and show that

$$\lim_{\delta \to 0}[U(f_\delta)](\xi) = -\pi i\mathrm{sgn}(\xi).$$

(We may say then that the Fourier transform of the non-integrable and non-square-integrable function $1/x$ is the function $-\pi i\mathrm{sgn}$.)

EXERCISE 6.16. (Hausdorff-Young Inequality) Suppose $f \in L^1 \cap L^p$ for $1 \leq p \leq 2$. Prove that $\hat{f} \in L^{p'}$, for $1/p + 1/p' = 1$, and that $\|\hat{f}\|_{p'} \leq \|f\|_p$. HINT: Use the Riesz Interpolation Theorem.

DEFINITION. If u is a tempered distribution, i.e., an element of $\mathcal{S}'$, define the *Fourier transform* $\hat{u}$ of u to be the linear functional on $\mathcal{S}$ given by

$$\hat{u}(f) = u(\hat{f}).$$

EXERCISE 6.17. (a) Prove that the Fourier transform of a tempered distribution is again a tempered distribution.

(b) Suppose h is a tempered function in $L^1(\mathbb{R})$ $(L^2(\mathbb{R}))$, and suppose that u is the tempered distribution u_h. Show that $\hat{u} = u_{\hat{h}}$ $(u_{U(h)})$.

(c) If u is the tempered distribution defined by

$$u(f) = \lim_{\delta \to 0} \int_{|t| \geq \delta} [f(t)/t]\, dt,$$

show that $\hat{u} = u_{-\pi i \mathrm{sgn}}$. See part b of Exercise 5.8.

(d) If u is a tempered distribution, show that the Fourier transform of the tempered distribution u' is the tempered distribution $v = m\hat{u}$, where m is the C^∞ tempered function given by $m(\xi) = 2\pi i \xi$. That is,

$$\widehat{u'}(f) = v(f) = \hat{u}(mf).$$

(e) Suppose U and its distributional derivative u' are both tempered distributions corresponding to L^2 functions f and g respectively. Prove that f is absolutely continuous and that $f'(x) = -g(x)$ a.e.

(f) Suppose both u and its distributional derivative u' are tempered distributions corresponding to L^2 functions f and g respectively. Assume that there exists an $\epsilon > 0$ such that $|\xi|^{(3/2)+\epsilon}\hat{u}(\xi)$ is in $L^2(\mathbb{R})$. Prove that f is in fact a C^1 function.

DEFINITION. For vectors x and y in $\mathbb{R}^n$, write (x, y) for the dot product of x and y. If $f \in L^1(\mathbb{R}^n)$, define the *Fourier transform* $\hat{f}$ of f on $\mathbb{R}^n$ by

$$\hat{f}(\xi) = \int_{\mathbb{R}^n} f(x)e^{-2\pi i(x,\xi)}\, dx.$$

EXERCISE 6.18. (a) Prove the Riemann-Lebesgue theorem: If $f \in L^1(\mathbb{R}^n)$, then $\hat{f} \in C_0(\mathbb{R}^n)$.

(b) Prove the convolution theorem: If $f, g \in L^1(\mathbb{R}^n)$, then

$$\widehat{f * g} = \hat{f}\hat{g}.$$

Show also that $\int \hat{f}g = \int f\hat{g}$.

(c) Let $1 \leq i \leq n$, and suppose that both f and its partial derivative $\dfrac{\partial f}{\partial x_i}$ belong to $L^1(\mathbb{R}^n)$. Show that

$$\widehat{\frac{\partial f}{\partial x_i}}(\xi) = -2\pi i \xi_i \hat{f}(\xi).$$

Generalize this equality to higher order and mixed partial derivatives.

(d) For $t > 0$ define g_t on $\mathbb{R}^n$ by

$$g_t(x) = (2\pi t)^{-n/2} e^{-\|x\|^2/2t}.$$

Show that

$$\hat{g}_t(\xi) = e^{-2\pi t \|\xi\|^2}.$$

(e) Prove the Inversion Theorem for the Fourier transform on $L^1(\mathbb{R}^n)$; i.e., if $f, \hat{f} \in L^1(\mathbb{R}^n)$, show that

$$f(x) = \int_{\mathbb{R}^n} \hat{f}(\xi) e^{2\pi i(x,\xi)} \, d\xi$$

for almost all $x \in \mathbb{R}^n$.

(f) Prove the Plancherel Formula for the Fourier transform on $L^2(\mathbb{R}^n)$; i.e., for $f, g \in L^1(\mathbb{R}^n) \cap L^2(\mathbb{R}^n)$, show that

$$\int f\bar{g} = \int \hat{f}\bar{\hat{g}}.$$

Verify that the Fourier transform has a unique extension from $L^1(\mathbb{R}^n) \cap L^2(\mathbb{R}^n)$ to an isometry from $L^2(\mathbb{R}^n)$ onto itself. We denote this isometry by U and call it the L^2 *Fourier transform* on $\mathbb{R}^n$.

EXERCISE 6.19. (The Green's Function for the Laplacian) Let L denote the Laplacian on $\mathbb{R}^n$; i.e.,

$$L(u) = \sum_{i=1}^{n} \frac{\partial^2 u}{\partial x_i{}^2},$$

for u any almost everywhere twice differentiable function on $\mathbb{R}^n$. Let D be the space of all functions $u \in L^2(\mathbb{R}^n)$, all of whose first and second order partial derivatives are continuous and belong to $L^2(\mathbb{R}^n)$. Think of L as a mapping of D into $L^2(\mathbb{R}^n)$. Let $\tilde{D}$ be the set of all $f \in L^2(\mathbb{R}^n)$ such that $\|\xi\|^2 U(f)(\xi)$ belongs to $L^2(\mathbb{R}^n)$, and define $\tilde{L} : \tilde{D} \to L^2(\mathbb{R}^n)$ by $\tilde{L}(f) = U^{-1}(mU(f))$, where U denotes the L^2 Fourier transform on $\mathbb{R}^n$, and m is the function defined by $m(\xi) = -4\pi^2 \|\xi\|^2$.

(a) Show that $D \subseteq \tilde{D}$, and that the graph of $\tilde{L}$ is closed.

(b) Assume that $n \geq 3$. Find a Green's function g for $\tilde{L}$, and observe that g is also a Green's function for L. HINT: Set

$$p(x) = c/(1 + \|x\|^2)^{(n+1)/2},$$

find a $u_0 \in D$ such that $L(u_0) = \tilde{L}(u_0) = p$. Then use Exercises 6.7 and 6.12.

(c) Find a Green's function for the Laplacian in $\mathbb{R}^2$ and in $\mathbb{R}$.

Hilbert Transform on the Line

If m is a bounded measurable function on $\mathbb{R}$, we may define a bounded operator M on L^2 by

$$M(f) = U^{-1}(mU(f)),$$

where U denotes the L^2 Fourier transform. Such an operator M is called a *multiplier operator* or simply a *multiplier*.

EXERCISE 6.20. Suppose $m = \hat{f}$ for some L^1 function f. Show that the multiplier operator M is given by

$$M(g) = f * g.$$

Note, therefore, that multipliers are generalizations of L^1 convolution operators.

REMARK. Recall from Theorem 6.2 that L^1 convolution operators determine bounded operators on every L^p space ($1 \le p \le \infty$). If m is not the Fourier transform of an L^1 function, then the multiplier M (a priori a bounded operator on $L^2(\mathbb{R})$) may or may not have extensions to bounded operators on L^p spaces other than $p = 2$, and it is frequently important to know when it does have such extensions.

EXERCISE 6.21. Let m be a bounded measurable function on $\mathbb{R}$.

(a) Suppose that the multiplier M, corresponding to the function m, determines a bounded operator from $L^p(\mathbb{R})$ into itself for every $1 < p < \infty$. Show that the multiplier corresponding to the function $\overline{m}$ is the adjoint M^* of M, and hence is a bounded operator from $L^q(\mathbb{R})$ into itself for every $1 < q < \infty$.

(b) Prove that the multiplier M, corresponding to the function m, determines a bounded operator from $L^p(\mathbb{R})$ into itself, for some $1 < p < \infty$, if and only if M is a bounded operator from $L^{p'}(\mathbb{R})$ into itself, where $1/p + 1/p' = 1$.

Perhaps the most important example of a nontrivial multiplier is the following.

DEFINITION. Let h denote the function $-i\mathrm{sgn}$, where sgn is the signum function. The *Hilbert transform* is the multiplier operator H corresponding to the function h; i.e., on $L^2(\mathbb{R})$ we have

$$H(f) = U^{-1}(-i\mathrm{sgn}U(f)).$$

REMARK. In view of the results in Exercises 6.15 and 6.20, we might expect the Hilbert transform to correspond somehow to convolution by the nonintegrable function $1/\pi x$. Indeed, this is what we shall see below.

EXERCISE 6.22. (a) Show that the Hilbert transform has no extension to a bounded operator on L^1. HINT: For $f \in L^1(\mathbb{R}) \cap L^2(\mathbb{R})$, we have that $U(f) = \hat{f}$ is continuous.

(b) Suppose $f \in L^1(\mathbb{R}) \cap L^2(\mathbb{R})$, and $\hat{f} \in L^1(\mathbb{R})$. Verify the following sequence of equalities:

$$
\begin{aligned}
[H(f)](x) &= [U^{-1}(-i\mathrm{sgn}\hat{f})](x) \\
&= (1/\pi) \lim_{\delta \to 0}[U^{-1}(U(f_\delta)\hat{f})](x) \\
&= (1/\pi) \lim_{\delta \to 0} \int_{-\infty}^{\infty} \hat{f}(\xi)[U(f_\delta)](\xi)e^{2\pi i x\xi}\, d\xi \\
&= (1/\pi) \lim_{\delta \to 0} \int_{-\infty}^{\infty} f(x+t)f_\delta^*(t)\, dt \\
&= \lim_{\delta \to 0} \int_{-\infty}^{\infty} f(x-t)/\pi t\, dt,
\end{aligned}
$$

where f_δ is the function from part e of Exercise 6.15. Note that this shows that the operator H can be thought of as a generalization of convolution, in this case by the nonintegrable function $1/\pi x$.

(c) Verify that if f is a real-valued function in $L^1(\mathbb{R}) \cap L^2(\mathbb{R})$, and if $\hat{f}$ is in $L^1(\mathbb{R})$, then $H(f)$ also is real-valued.

EXERCISE 6.23. (a) For each positive real number y, define the function g_y by

$$g_y(\xi) = e^{-2\pi y|\xi|}.$$

Show that

$$\mathrm{sgn}(\xi)g_y(\xi) = (-1/2\pi y)g_y'(\xi)$$

for every $y > 0$ and every $\xi \neq 0$.

(b) Let f be a Schwartz function. For any real x, let f_x denote the function defined by

$$f_x(y) = f(x + y).$$

Verify the following sequence of equalities:

$$[H(f)](x) = \lim_{y \to 0} (i/2\pi y)[U^{-1}(g_y' \hat{f})](x)$$

$$= \lim_{y \to 0} (i/2\pi y) \int_{-\infty}^{\infty} g_y'(\xi) \hat{f}(\xi) e^{2\pi i x \xi} \, d\xi$$

$$= \lim_{y \to 0} (-i/2\pi y) \int_{-\infty}^{\infty} g_y(\xi) \hat{f_x}'(\xi) \, d\xi$$

$$= \lim_{y \to 0} (-i/2\pi y) \int_{-\infty}^{\infty} \hat{g_y}(t)(-2\pi i t) f_x(t) \, dt$$

$$= \lim_{y \to 0} \int_{-\infty}^{\infty} \frac{t/\pi}{t^2 + y^2} f(x - t) \, dt.$$

Note again that the Hilbert transform can be regarded as a kind of convolution by $1/\pi x$.

THEOREM 6.9. *The Hilbert transform determines a bounded operator from L^p into itself, for each $1 < p < \infty$.*

PROOF. Given a $1 < p < \infty$, it will suffice to prove that there exists a positive constant c_p such that

$$\|H(f)\|_p \le c_p \|f\|_p$$

for all real-valued, C^∞ functions f having compact support. (Why?) First, let n be a positive integer, and let $p = 2n$. For such a fixed real-valued, C^∞ function f having compact support, define a function F of a complex variable $z = x + iy$ by

$$F(z) = (1/\pi i) \int_{-\infty}^{\infty} f(t)/(t - z) \, dt.$$

Then F is analytic at each point $z = x + iy$ for $y > 0$ (it has a derivative there). It follows easily that there exists a constant c for which

$$|F(x + iy)| \le c/y \tag{6.4}$$

for all x and all $y > 0$, and

$$|F(x + iy)| \le c/|x| \tag{6.5}$$

for all $y > 0$ and all sufficiently large x. (See Exercise 6.24 below.)

If we write $F = U + iV$, then since f is real-valued we have

$$U(x + iy) = \int_{-\infty}^{\infty} \frac{y/\pi}{y^2 + (x - t)^2} f(t)\, dt$$

and

$$V(x + iy) = \int_{-\infty}^{\infty} \frac{(x - t)/\pi}{(x - t)^2 + y^2} f(t)\, dt.$$

Then, by Exercises 6.7 and 6.23, we have that for every real x

$$f(x) = \lim_{y \to 0} U(x + iy),$$

and

$$[H(f)](x) = \lim_{y \to 0} V(x + iy).$$

We fix a sequence $\{y_j\}$ converging to 0 and define $U_j(x) = U(x + iy_j)$ and $V_j(x) = V(x + iy_j)$. Then $f = \lim U_j$ and $H(f) = \lim V_j$.

Because F is analytic in the upper half plane, and because of inequalities (6.4) and (6.5), we have that

$$\int_{-\infty}^{\infty} F^{2n}(x + iy)\, dx = 0 \qquad (6.6)$$

for each positive y. (See Exercise 6.24.) Hence

$$\int_{-\infty}^{\infty} \Re(F^{2n})(x + iy)\, dx = 0$$

for every positive y.

From trigonometry, we see that there exist positive constants a_n and b_n such that

$$\sin^{2n}(\theta) \le a_n \cos^{2n}(\theta) + (-1)^n b_n \cos(2n\theta)$$

for all real θ. Indeed, choose b_n so that this is true for θ near $\pi/2$ and then choose a_n so the inequality holds for other θ's. It follows then that for any complex number z we have

$$\Im(z)^{2n} \le a_n \Re(z)^{2n} + (-1)^n b_n \Re(z^{2n}).$$

So, we have that

$$V(x+iy)^{2n} \leq a_n U(x+iy)^{2n} + (-1)^n b_n \Re(F^{2n}(x+iy)),$$

whence

$$\int_{-\infty}^{\infty} V(x+iy)^{2n}\, dx \leq a_n \int_{-\infty}^{\infty} U(x+iy)^{2n}\, dx$$

implying that

$$\int_{-\infty}^{\infty} |V_j(x)|^{2n}\, dx \leq a_n \int_{-\infty}^{\infty} |U_j(x)|^{2n}\, dx$$

for all j. So, by the dominated convergence theorem and part b of Exercise 6.7,

$$\int_{-\infty}^{\infty} |[H(f)](x)|^p\, dx \leq a_n \int_{-\infty}^{\infty} |f(x)|^p\, dx,$$

and

$$\|H(f)\|_p \leq a_n^{1/p} \|f\|_p,$$

where $p = 2n$.

We have thus shown that the Hilbert transform determines a bounded operator from L^p into itself, for p of the form $2n$. By the Riesz Interpolation Theorem, it follows then that the Hilbert transform determines a bounded operator from L^p into itself, for $2 \leq p < \infty$. The proof can now be completed by appealing to Exercise 6.21 for the cases $1 < p < 2$.

EXERCISE 6.24. (a) Show that any constant $c \geq \int |f(t)|\, dt$ will satisfy inequality (6.4). Supposing that f is supported in the interval $[-a, a]$, show that any constant $c \geq 2\int |f(t)|\, dt$ will satisfy inequality (6.5) if $|x| \geq 2a$.

(b) Establish Equation (6.6) by integrating around a large square contour.

(c) Let m be the characteristic function of an open interval (a, b), where $-\infty \leq a < b \leq \infty$. Prove that the multiplier M, corresponding to m determines a bounded operator on every L^p for $1 < p < \infty$. HINT: Write m as a finite linear combination of translates of $-i\operatorname{sgn}$.

(d) Let m be the characteristic function of the set $E = \cup_n [2n, 2n + 1]$. Verify that the multiplier M corresponding to m has no bounded extension to any L^p space for $p \neq 2$.

CHAPTER VII

AXIOMS FOR A MATHEMATICAL MODEL
OF EXPERIMENTAL SCIENCE

This chapter is a diversion from the main subject of this book, and it can be skipped without affecting the material that follows. However, we believe that the naive approach taken in this chapter toward the axiomatizing of experimental science serves as a good motivation for the mathematical theory developed in the following four chapters.

We describe here a set of axioms, first introduced by G.W. Mackey, to model experimental investigation of a system in nature. We suppose that we are studying a phenomenon in terms of various observations of it that we might make. We postulate that there exists a nonempty set S of what we shall call the possible *states* of the system, and we postulate that there is a nonempty set O of what we shall call the possible *observables* of the system. We give two examples.

(1) Suppose we are investigating a system that consists of a single physical particle in motion on an infinite straight line. Newtonian mechanics $(f = ma)$ tells us that the system is completely determined for all future time by the current position and velocity, i.e., by two real numbers. Hence, the states of this system might well be identified with points in the plane. Two of the (many) possible observables of this system can be described as position and velocity observables. We imagine that there is a device which indicates where the particle is and another device that indicates its velocity. More realistically, we might have many yes/no devices that answer the observational questions: "Is the particle between a and b?" "Is the velocity of the particle between c and d?"

Quantum mechanical models of this single particle are different from the Newtonian one. They begin by assuming that the (pure) states of this one-particle system are identifiable with certain square-integrable functions and the observables are identified with certain linear transformations. This model seems quite mysterious to most mathematicians, and Mackey's axioms form one attempt at justifying it.

(2) Next, let us imagine that we are investigating a system in which three electrical circuits are in a black box and are open or closed according to some process of which we are not certain. The states of this system might well be described as all triples of 0's and 1's (0 for open and 1 for closed). Suppose that we have only the following four devices for observing this system. First, we can press a button b_0 and determine how many of the three circuits are closed. However, when we press this button, it has the effect of opening all three circuits, so that we have no hope of learning exactly which of the three were closed. (Making the observation actually affects the system.) In addition, we have three other buttons b_1, b_2, b_3, b_i telling whether circuit i is open or closed. Again, when we press button b_i, all three circuits are opened, so that we have no way of determining if any of the circuits other than the ith was closed. This is a simple example in which certain simultaneous observations appear to be impossible, e.g., determining whether circuits 1 and 2 are both closed.

The axioms we introduce are concerned with the concept of interpreting what it means to make a certain observation of the system when the system is in a given state. The result of such an observation should be a real number, with some probability, depending on the state and on the observable.

AXIOM 1.   To each state $\alpha \in S$ and observable $A \in O$ there corresponds a Borel probability measure $\mu_{\alpha,A}$ on $\mathbb{R}$.

REMARK. The probability measure $\mu_{\alpha,A}$ contains the information about the probability that the observation A will result in a certain value, when the system is in the state α.

EXERCISE 7.1. Write out in words, from probability theory, what the following symbols mean.

(a) $\mu_{\alpha,A}([3,5]) = 0.9$.

(b) $\mu_{\alpha,A}(\{0\}) = 1$.

AXIOM 2.   (a) If A, B are observables for which $\mu_{\alpha,A} = \mu_{\alpha,B}$ for every state $\alpha \in S$, then $A = B$.

(b) If α, β are states for which $\mu_{\alpha,A} = \mu_{\beta,A}$ for every observable $A \in O$, then $\alpha = \beta$.

EXERCISE 7.2. Discuss the intuitive legitimacy of Axiom 2.

AXIOM 3. If $\alpha_1, \ldots \alpha_n$ are states, and $t_1, \ldots t_n$ are nonnegative real numbers for which $\sum_{i=1}^n t_i = 1$, then there exists a state α for which

$$\mu_{\alpha,A} = \sum_{i=1}^n t_i \mu_{\alpha_i,A}$$

for every observable A. This axiom can be interpreted as asserting that the set S of states is closed under convex combinations. If the α_i's are not all identical, we call this state α a *mixed state* and we write $\alpha = \sum_{i=1}^n t_i \alpha_i$.

We say that a state $\alpha \in S$ is a *pure state* if it is not a mixture of other states. That is, if $\alpha = \sum_{i=1}^n t_i \alpha_i$, with each $t_i > 0$ and $\sum_{i=1}^n t_i = 1$, then $\alpha_i = \alpha$ for all i.

EXERCISE 7.3. Discuss the intuitive legitimacy of Axiom 3. Think of a physical system, like a beaker of water, for which there are what we can interpret as pure states and mixed states.

AXIOM 4. If A is an observable, and $f : \mathbb{R} \to \mathbb{R}$ is a Borel function, then there exists an observable B such that

$$\mu_{\alpha,B}(E) = \mu_{\alpha,A}(f^{-1}(E))$$

for every state α and every Borel set $E \subseteq \mathbb{R}$. We denote this observable B by $f(A)$.

EXERCISE 7.4. Discuss the intuitive legitimacy of Axiom 4. Show that, when the system is in the state α and the observable A results in a value t with probability p, the observable $B = f(A)$ results in the value $f(t)$ with the same probability p.

EXERCISE 7.5. (a) Prove that there exists an observable A such that $\mu_{\alpha,A}(-\infty, 0) = 0$ for every state α. That is, A is an observable that is nonnegative with probability 1 independent of the state of the system. HINT: Use $f(t) = t^2$ for example.

(b) Given a real number t, show that there exists an observable A such that $\mu_{\alpha,A} = \delta_t$ for every state α. That is, A is an observable that equals t with probability 1, independent of the state of the system.

(c) Show that the set of observables is closed under scalar multiplication. That is, if A is an observable and c is a nonzero real number, then there exists an observable B such that

$$\mu_{\alpha,B}(E) = \mu_{\alpha,A}((1/c)E).$$

We may then write $B = cA$.

(d) If A and B are observables, does there have to be an observable C that we could think of as the sum $A + B$?

(e) In what way must we alter the descriptions of the systems in Example 1 and Example 2 in order to incorporate these first four axioms (particularly Axioms 3 and 4)?

DEFINITION. We say that two observables A and B are *compatible, pairwise compatible,* or *simultaneously observable* if there exists an observable C and Borel functions f and g such that $A = f(C)$ and $B = g(C)$. A sequence $\{A_i\}$ is called *mutually compatible* if there exists an observable C and Borel functions $\{f_i\}$ such that $A_i = f_i(C)$ for all i.

EXERCISE 7.6. Is there a difference between a sequence $\{A_i\}$ of observables being pairwise compatible and being mutually compatible? In particular, is it possible that there could exist observables A, B, C, such that A and B are compatible, B and C are compatible, A and C are compatible, and yet A, B, C are not mutually compatible? HINT: Try to modify Example 2.

EXERCISE 7.7. (a) If A, B are observables, what should it mean to say that an observable C is the sum $A + B$ of A and B? Discuss why we do not hypothesize that there always exists such an observable C.

(b) If A and B are compatible, can we prove that there exists an observable C that can be regarded as $A + B$?

DEFINITION. An observable q is called a *question* or a *yes/no observable* if, for each state α, the measure $\mu_{\alpha,q}$ is supported on the two numbers 0 and 1. We say that the result of observing q, when the system is in the state α, is "yes" with probability $\mu_{\alpha,q}(\{1\})$, and it is "no" with probability $\mu_{\alpha,q}(\{0\})$.

THEOREM 7.1. *Let A be an observable.*

(1) *For each Borel subset E in $\mathbb{R}$, the observable $\chi_E(A)$ is a question.*

(2) *If g is a real-valued Borel function on $\mathbb{R}$, for which $g(A)$ is a question, then there exists a Borel set E such that $g(A) = \chi_E(A)$.*

(Note that condition 2 does not assert that g necessarily equals χ_E.)

PROOF. For each Borel set E, we have

$$\mu_{\alpha,\chi_E(A)}(\{1\}) = \mu_{\alpha,A}(\chi_E^{-1}(\{1\}))$$
$$= \mu_{\alpha,A}(E),$$

and

$$\mu_{\alpha,\chi_E(A)}(\{0\}) = \mu_{\alpha,A}(\chi_E^{-1}(\{0\}))$$
$$= \mu_{\alpha,A}(\tilde{E})$$
$$= 1 - \mu_{\alpha,A}(E),$$

which proves that $\mu_{\alpha,\chi_E(A)}$ is supported on the two points 0 and 1 for every α, whence $\chi_E(A)$ is a question and so part 1 is proved.

Given a g for which $q = g(A)$ is a question, set $E = g^{-1}(\{1\})$, and observe that for any $\alpha \in S$ we have

$$\mu_{\alpha,q}(\{1\}) = \mu_{\alpha,g(A)}(\{1\})$$
$$= \mu_{\alpha,A}(E)$$
$$= \mu_{\alpha,\chi_E(A)}(\{1\}).$$

Since both q and $\chi_E(A)$ are questions, it follows from the preceding paragraph that

$$\mu_{\alpha,q}(\{0\}) = \mu_{\alpha,\chi_E(A)}(\{0\}),$$

showing that

$$\mu_{\alpha,q} = \mu_{\alpha,\chi_E(A)}$$

for every state α. Then, by Axiom 2, we have that

$$g(A) = q = \chi_E(A).$$

We now define some mathematical structure on the set Q of all questions. This set will form the fundamental ingredient of our model.

DEFINITION. Let Q denote the set of all questions. For each question $q \in Q$, define a real-valued function m_q on the set S of states by

$$m_q(\alpha) = \mu_{\alpha,q}(\{1\}).$$

If p and q are questions, we say that $p \leq q$ if $m_p(\alpha) \leq m_q(\alpha)$ for all $\alpha \in S$.

If p, q and r are questions, for which $m_r = m_p + m_q$, we say that p and q are *summable* and that r is the sum of p and q. We then write $r = p + q$. More generally, if $\{q_i\}$ is a countable (finite or infinite) set of questions, we say that the q_i's are *summable* if there exists a question q such that

$$m_q(\alpha) = \sum_i m_{q_i}(\alpha)$$

for every $\alpha \in S$. In such a case, we write $q = \sum_i q_i$.

Finally, a countable set $\{q_i\}$ is called *mutually summable* if every subset of the q_i's is summable.

REMARK. As mentioned above, the set Q will turn out to be the fundamental ingredient of our model, in the sense that everything else will be described in terms of Q.

THEOREM 7.2.

(1) *The set Q is a partially ordered set with respect to the ordering $\leq$ defined above.*

(2) *There exists a question $q_1 \in Q$, which we shall often simply call 1, for which $q \leq q_1$ for every $q \in Q$. That is, Q has a maximum element q_1.*

(3) *There exists a question $q_0 \in Q$, which we shall often simply call 0, for which $q_0 \leq q$ for every $q \in Q$. That is, Q has a minimum element q_0.*

(4) *For each question q, there exists a question $\tilde{q}$ such that*

$$m_q + m_{\tilde{q}} = q_1 = 1.$$

That is, every question has a complementary question.

PROOF. That Q is a partially ordered set is clear.

If A is any observable, and f is the identically 1 function, then the question $q_1 = f(A)$ satisfies

$$
\begin{aligned}
m_{q_1}(\alpha) &= \mu_{\alpha, q_1}(\{1\}) \\
&= \mu_{\alpha, f(A)}(\{1\}) \\
&= \mu_{\alpha, A}(f^{-1}(\{1\})) \\
&= \mu_{\alpha, A}(\mathbb{R}) \\
&= 1
\end{aligned}
$$

for all α, and clearly then $q \leq q_1$ for every $q \in Q$.

Taking f to be the identically 0 function, we may define the question q_0 to be $f(A)$.

Finally, if f is the function defined by $f(t) = 1 - t$, and if $q \in Q$, then $f(q)$ is the desired question $\tilde{q}$. Indeed,

$$\mu_{\alpha, f(q)}(\{1\}) = \mu_{\alpha, q}(f^{-1}(\{1\}))$$
$$= \mu_{\alpha, q}(\{0\}),$$

and

$$\mu_{\alpha, f(q)}(\{0\}) = \mu_{\alpha, q}(f^{-1}(\{0\}))$$
$$= \mu_{\alpha, q}(\{1\}),$$

proving that $f(q)$ is a question and showing also that

$$m_{f(q)}(\alpha) = 1 - m_q(\alpha)$$

for every α, as desired.

DEFINITION. Two questions p and q are called *orthogonal* if $p \le \tilde{q}$ or (equivalently) $q \le \tilde{p}$. That is, p and q are orthogonal if $m_p + m_q \le 1$.

REMARK. Clearly, if p and q are summable, then they are orthogonal, but the converse need not hold. Even if $m_p + m_q \le 1$, there may not be a question r such that $m_r = m_p + m_q$. We have no axiom that ensures this.

Our next goal is to describe the observables in terms of the set Q.

THEOREM 7.3. *Let A be an observable. For each Borel set $E \subseteq \mathbb{R}$, put*

$$q_E^A = \chi_E(A).$$

Then, the mapping $E \to q_E^A$ satisfies:

(1) $q_{\mathbb{R}}^A = 1$ and $q_{\emptyset}^A = 0$.

(2) *If $\{E_i\}$ is a sequence of pairwise disjoint Borel sets, then $\{q_{E_i}^A\}$ is a sequence of mutually compatible, mutually summable, (pairwise orthogonal) questions, and*

$$q_{\cup_i E_i}^A = \sum_i q_{E_i}^A.$$

(3) *If A and B are observables, for which $q_E^A = q_E^B$ for every Borel set E, then $A = B$.*

PROOF. Since $\chi_\mathbb{R}$ is the identically 1 function, it follows that $q^A_\mathbb{R} = 1$. Similarly, $q^A_\emptyset = 0$.

If $\{F_i\}$ is any (finite or infinite) sequence of pairwise disjoint Borel sets, set $F = \cup F_i$. Then, clearly the questions $\{q^A_{F_i}\}$ are mutually compatible, since they are all functions of the observable A. Also, for any state α we have

$$
\begin{aligned}
m_{q^A_F}(\alpha) &= \mu_{\alpha, q^A_F}(\{1\}) \\
&= \mu_{\alpha, \chi_F(A)}(\{1\}) \\
&= \mu_{\alpha, A}(F) \\
&= \mu_{\alpha, A}(\cup F_i) \\
&= \sum_i \mu_{\alpha, A}(F_i) \\
&= \sum_i \mu_{\alpha, \chi_{F_i}(A)}(\{1\}) \\
&= \sum_i m_{q^A_{F_i}}(\alpha).
\end{aligned}
$$

Now let $\{E_i\}$ be a sequence of pairwise disjoint Borel sets. The preceding calculation, as applied to every subset of the E_i's, shows that the questions $\{q^A_{E_i}\}$ are mutually summable and that

$$
q^A_E = \sum q^A_{E_i}.
$$

And, in particular, since the $q^A_{E_i}$'s are pairwise summable, they are pairwise orthogonal.

Finally, if A and B are distinct observables, then, by Axiom 2, there exists a state α such that $\mu_{\alpha, A} \neq \mu_{\alpha, B}$. Hence, there is a Borel set E such that

$$
\mu_{\alpha, A}(E) \neq \mu_{\alpha, B}(E),
$$

or

$$
\mu_{\alpha, \chi_E(A)}(\{1\}) \neq \mu_{\alpha, \chi_E(B)}(\{1\}),
$$

or $q^A_E \neq q^B_E$, as desired.

DEFINITION. A mapping $E \rightarrow q_E$, from the σ-algebra $\mathcal{B}$ of Borel sets into Q, which satisfies the two properties below, is called a *question-valued measure*.

(1) $q_\mathbb{R} = 1$ and $q_\emptyset = 0$.

(2) If $\{E_i\}$ is a sequence of pairwise disjoint Borel sets, then $\{q_{E_i}\}$ is a sequence of mutually compatible, mutually summable, (pairwise orthogonal) questions, and

$$q_{\cup E_i} = \sum_i q_{E_i}.$$

REMARK. Theorem 7.3 asserts that each observable A determines a question-valued measure q^A and that the assignment $A \to q^A$ is 1-1.

EXERCISE 7.8. Let $E \to q_E$ be a question-valued measure.
(a) Prove that if $E \subseteq F$, then $q_E \leq q_F$; i.e., $E \to q_E$ is order-preserving.
(b) Show that $q_{\tilde{E}} = \tilde{q_E}$; i.e., $E \to q_E$ is complement-preserving.

AXIOM 5. If $E \to q_E$ is a question-valued measure, then there exists an observable A such that $q_E = q_E^A$ for all Borel sets E.

EXERCISE 7.9. Discuss the intuitive legitimacy of Axiom 5.

EXERCISE 7.10. Let $\{q_1, q_2, \ldots\}$ be a mutually summable set of questions for which $\sum_i q_i = 1$. Prove that the q_i's are mutually com patible. HINT: Define a question-valued measure $E \to q_E$ by setting $q_{\{i\}} = q_i$ for each $i = 1, 2, \ldots$, and define

$$q_E = \sum_{i \in E} q_{\{i\}}.$$

then use Axiom 5.

THEOREM 7.4. *Let p and q be questions. Then p and q are compatible if and only if there exist mutually summable questions r_1, r_2, r_3 and r_4 satisfying:*
(1) $p = r_1 + r_2$.
(2) $q = r_2 + r_3$.
(3) $r_1 + r_2 + r_3 + r_4 = 1$.

PROOF. If p and q are compatible, let A be an observable and let f and g be Borel functions such that $p = f(A)$ and $q = g(A)$. By Theorem 7.1, we may assume that $f = \chi_E$ and $g = \chi_F$, where E and F are Borel sets in $\mathbb{R}$. Define four pairwise disjoint Borel sets as follows:

$$E_1 = E - F, \; E_2 = E \cap F, \; E_3 = F - E, \; E_4 = \mathbb{R} - (E \cup F).$$

Now, define $r_i = \chi_{E_i}(A)$. The desired properties of the r_i's follow directly. For example,

$$
\begin{aligned}
m_p(\alpha) &= \mu_{\alpha, \chi_E(A)}(\{1\}) \\
&= \mu_{\alpha, A}(E) \\
&= \mu_{\alpha, A}(E_1 \cup E_2) \\
&= \mu_{\alpha, A}(E_1) + \mu_{\alpha, A}(E_2) \\
&= \mu_{\alpha, \chi_{E_1}(A)}(\{1\}) + \mu_{\alpha, \chi_{E_2}(A)}(\{1\}) \\
&= m_{r_1}(\alpha) + m_{r_2}(\alpha),
\end{aligned}
$$

showing that $p = r_1 + r_2$ as desired. We leave the other verifications to the exercise that follows.

Conversely, given $r_1, \ldots, r_4$ satisfying the conditions in the statement of the theorem, define a mapping $E \to q_E$ of the σ-algebra $\mathcal{B}$ of Borel sets into Q as follows:

$$
q_E = \sum_{i \in E} r_i,
$$

with the convention that $q_E = 0$ if E does not contain any of the numbers 1,2,3,4. Then $E \to q_E$ is a question-valued measure. (See the preceding exercise.) By Axiom 5, there exists an observable A such that $q_E = q_E^A$ for all E, and clearly $p = \chi_{[1,2]}(A)$ and $q = \chi_{[2,3]}(A)$ are both functions of A, as desired.

EXERCISE 7.11. Verify that $q = r_1 + r_3$ and that $r_1 + r_2 + r_3 + r_4 = 1$ in the first part of the preceding proof.

EXERCISE 7.12. (a) Prove that the map $q \to m_q$ is 1-1.

(b) Show, by identifying Q with m_q, that the set Q can be given a natural Hausdorff topology.

(c) Let q be a question. Show that the set of all questions p, for which $p \leq q$, and the set of all questions p such that p is orthogonal to q are closed subsets of Q in the topology from part b.

(d) Prove that the map $q \to \tilde{q}$ is continuous with respect to the topology on Q from part b.

REMARK. We equip the set Q of all questions with the topology from the preceding exercise. That is, we identify each question q with the corresponding function m_q and use the topology of pointwise convergence of these functions. In this way, the set Q is a partially-ordered Hausdorff topological space having a maximum element and a minimum element. In addition to these topological and order structures on Q, there are notions of complement, of orthogonality, of summability, and of compatibility. We shall be interested in finding a mathematical object having these attributes.

EXERCISE 7.13. (a) Show that the closed interval $[0, 1]$ has all the properties of Q. That is, show that $[0, 1]$ is a partially-ordered topological space having a maximum and a minimum, and show that there is a notion of summability (not the usual one) on $[0, 1]$ such that each element has a complement. Finally, prove that any two elements of $[0, 1]$ that are summable are compatible. In a way, $[0, 1]$ is the simplest model for Q. HINT: Use the characterization of compatibility in Theorem 7.4.

(b) Is the unit circle a possible model for Q?

Having described the set O of observables as question-valued measures, we turn next to the set S of states. We want to describe the states also in terms of the set Q.

DEFINITION. By an *automorphism* of Q we mean a 1-1 map ϕ of Q onto itself that satisfies:

(1) If $p \leq q$, then $\phi(p) \leq \phi(q)$; i.e., ϕ is order-preserving.
(2) $\phi(\tilde{q}) = \widetilde{\phi(q)}$ for all $q \in Q$; i.e., ϕ is complement-preserving.
(3) If $\{q_i\}$ is a summable set of questions, then $\{\phi(q_i)\}$ is a summable set of questions, and

$$\phi(\sum_i q_i) = \sum_i \phi(q_i).$$

If ϕ and ϕ^{-1} are Borel maps of the topological space Q, then ϕ is called a *Borel automorphism*.

By a *character* of the set Q of questions, we mean a continuous function $\mu : Q \to [0, 1]$ that satisfies:

(1) If $p \leq q$, then $\mu(p) \leq \mu(q)$; i.e., μ is order-preserving.
(2) $\mu(\tilde{q}) = 1 - \mu(q)$; i.e., μ is complement-preserving.
(3) If $\{q_i\}$ is a summable sequence of questions, then $\mu(\sum q_i) = \sum \mu(q_i)$; i.e., μ is additive when possible.

DEFINITION. For each state α, define a function μ_α on Q by

$$\mu_\alpha(q) = m_q(\alpha) = \mu_{\alpha,q}(\{1\}).$$

EXERCISE 7.14. (a) Show that each function μ_α is a continuous order-preserving map of Q into $[0,1]$.

(b) Show that $\mu_\alpha(\tilde{q}) = 1 - \mu_\alpha(q)$ for all $q \in Q$ and all $\alpha \in S$.

(c) If $\{q_i\}$ is a summable sequence of questions with $q = \sum q_i$, show that

$$\mu_\alpha(q) = \sum \mu_\alpha(q_i).$$

(d) Conclude that each function μ_α is a continuous character of Q.

(e) Show that the composition of a character of Q (e.g., μ_α) and a question-valued measure $E \to q_E$ defines a probability measure on the Borel subsets of $\mathbb{R}$.

(f) Show that the map $\alpha \to \mu_\alpha$ is 1-1 on S. Show further that if α is a mixed state, say $\alpha = \sum_{i=1}^{n} t_i \alpha_i$, then

$$\mu_\alpha = \sum_{i=1}^{n} t_i \mu_{\alpha_i};$$

i.e., $\alpha \to \mu_\alpha$ is an affine map on S.

REMARK. We give to S the Hausdorff topology obtained by identifying α with the continuous function μ_α on Q and considering this space of functions as topologized by the topology of pointwise convergence. Thus, we identify the set S of states of our system with certain continuous functions (characters) from the set Q of questions into $[0, 1]$. Of course, not every continuous function $f : Q \to [0, 1]$ need correspond to a state. Indeed, the functions corresponding to states must be characters.

We turn now to the evolution of the system in time. The axiom we take assumes that the system has always existed and will always exist. That is, the system can be thought of as evolving backward in time as well as forward. See part d of Exercise 7.15.

AXIOM 6. (Time Evolution of the System) For each nonnegative real number t, there exists a 1-1 transformation ϕ_t of S onto itself that describes the evolution of the system in time. In addition, for each nonnegative real number t, there exists a corresponding 1-1 transformation ϕ'_t, of the set Q onto itself, so that

(1) $\phi_{t+s} = \phi_t \circ \phi_s$ for all nonnegative s, t.

(2) For all $\alpha \in S$, $q \in Q$, and $t \geq 0$, we have

$$\mu_{\phi_t(\alpha),q} = \mu_{\alpha,\phi'_t(q)}.$$

(3) The map $(t, \alpha) \to \phi_t(\alpha)$ is a Borel map of $[0, \infty) \times S$ into S.

(4) The map $(t, q) \to \phi'_t(q)$ is a Borel map of $[0, \infty) \times Q$ into Q.

EXERCISE 7.15. (a) Discuss the intuitive legitimacy of Axiom 6. In particular, what is the interpretation of the transformation ϕ'_t?

(b) Show that $\phi'_{t+s} = \phi'_t \circ \phi'_s$ for all nonnegative t and s.

(c) Show that ϕ'_t is uniquely determined by ϕ_t and that ϕ_t is uniquely determined by ϕ'_t.

(d) Suppose α is a state. Given $t > 0$, show that there exists a unique state β such that if the system is in the state α now, then it was in the state β t units of time ago. (In other words, the evolution of the system can be reversed in time.)

REMARK. Of course, the primary goal of experimental investigation is to discover how to predict what will happen to a system as time goes by. In our development, then, we would want to discover the evolution transformations ϕ_t of S into itself.

Next, we turn to the notion of a symmetry of the system.

DEFINITION. If g denotes a (possibly hypothetical) 1-1 transformation of space, of the observer, of the system, etc., and if $\alpha \in S$ and $A \in O$ are given, we write $\mu^g_{\alpha,A}$ for the probability measure obtained by assuming that this transformation g has been performed, supposing that the system is in the state α, and by making the observation A. The transformation g is called a *symmetry* of the system if each $\mu^g_{\alpha,A} = \mu_{\alpha,A}$, i.e., if the "measurements" of the system are unchanged by performing the transformation g.

REMARK. Clearly, the set G of all symmetries forms a group of transformations.

AXIOM 7. To each symmetry g of the system there corresponds a 1-1 transformation π_g of S onto itself and a 1-1 transformation π'_g of Q onto itself such that

(1) $\pi_{g_1 g_2} = \pi_{g_1} \circ \pi_{g_2}$ for all $g_1, g_2 \in G$.

(2) For all $\alpha \in S$ and all $q \in Q$, we have

$$\mu_{\pi_g(\alpha),q} = \mu_{\alpha,\pi'_g(q)}.$$

(3) If a subgroup H of the group of all symmetries has some "natural" topological structure, then the maps $(h, \alpha) \to \pi_h(\alpha)$ and $(h, q) \to \pi'_h(q)$ are Borel maps from $H \times S$ into S and $H \times Q$ into Q respectively.

(4) π_g commutes with each evolution transformation ϕ_t; i.e., $\pi_g \circ \phi_t = \phi_t \circ \pi_g$ for all $t \geq 0$ and all $g \in G$.

EXERCISE 7.16. (a) Discuss the intuitive legitimacy of Axiom 7. In particular, what is the interpretation of the assumption that each π_g commutes with each evolution transformation ϕ_t?

(b) Show that each π'_g is uniquely determined by π_g, and that $\pi'_{g_1 g_2} = \pi'_{g_1} \circ \pi'_{g_2}$ for all $g_1, g_2 \in G$.

(c) Prove that each transformation π'_g commutes with each evolution transformation ϕ'_t.

THEOREM 7.5. *Each of the time evolution transformations ϕ'_t and each of the symmetry transformations π'_g are Borel automorphisms of the set Q. That is,*

(1) ϕ'_t, π'_g, and their inverses are Borel maps of Q onto itself.

(2) if $p \le q$, then $\phi'_t(p) \le \phi'_t(q)$ and $\pi'_g(p) \le \pi'_g(q)$.

(3) $\phi'_t(\tilde{q}) = \widetilde{\phi'_t(q)}$ and $\pi'_g(\tilde{q}) = \widetilde{\pi'_g(q)}$.

(4) If $\{q_i\}$ is a summable sequence of questions, then $\{\phi'_t(q_i)\}$ and $\{\pi'_g(q_i)\}$ are summable sequences of questions, and

$$\phi'_t\left(\sum q_i\right) = \sum \phi'_t(q_i)$$

and

$$\pi'_g\left(\sum q_i\right) = \sum \pi'_g(q_i).$$

PROOF. Suppose $p \le q$ are questions. We have then for any α that

$$m_{\phi'_t(q)}(\alpha) = \mu_{\alpha, \phi'_t(q)}(\{1\})$$
$$= \mu_{\phi_t(\alpha), q}(\{1\})$$
$$= m_q(\phi_t(\alpha))$$
$$\ge m_p(\phi_t(\alpha))$$
$$= m_{\phi'_t(p)}(\alpha),$$

showing that $\phi'_t(q) \ge \phi'_t(p)$. An analogous computation shows that $\pi'_g(q) \ge \pi'_g(p)$.

We leave the rest of the proof to the exercise that follows.

EXERCISE 7.17. Complete the proof to the preceding theorem.

We summarize the ingredients in our model as follows:

(1) There exists a Hausdorff space Q that is a partially ordered set, having a maximum element 1 and a minimum element 0. There are

notions of compatibility, orthogonality, and summability for certain of the elements of Q. Compatibility is characterized in Theorem 7.4.

(2) Each $q \in Q$ has a complementary element $\tilde{q}$ satisfying $q + \tilde{q} = 1$.

(3) The set S of states is represented as a set of continuous homomorphisms (characters) μ of Q into $[0,1]$. Each of these homomorphisms is continuous, order-preserving, additive when possible, and complement-preserving. This set S of states is a topological space and is closed under convex combinations.

(4) The set O of observables is identified with the set of Q-valued measures.

(5) The time evolution of the system is described by a one-parameter semigroup ϕ_t' of Borel transformations (automorphisms) of Q. These transformations are additive when possible, complement-preserving, and order-preserving.

(6) To each symmetry g of the system there corresponds a 1-1 transformation (automorphism) π_g' of Q onto itself. The transformation π_g' is Borel, preserves order, addition when possible, and complements. Each symmetry transformation π_g' commutes with each evolution transformation ϕ_t'.

The goal is to find concrete mathematical examples of the objects Q, S, ϕ_t' and π_g'. Initially, we will select a model for Q, and this selection will depend very much on which particular system we are studying. The set S is then a subset of the characters of Q, which, in any particular case, we can hope to describe concretely. Of course, the ultimate aim is to determine the evolution transformations ϕ_t of S into itself. Sometimes it is possible to describe the symmetry transformations π_g' by using group theory. If so, we may be able to describe the evolution transformations ϕ_t' by examining what transformations commute with the concrete π_g''s we have. However, our first task is to find an appropriate model for Q, and this we do in the next chapter.

We mention next some possibly less intuitively acceptable axioms. From a mathematical point of view, however, they are technically simplifying.

AXIOM 8. If $\{\alpha_i\}$ is a sequence of states, and if $\{t_i\}$ is a sequence of positive real numbers for which $\sum t_i = 1$, then there exists a state α, which we denote by $\sum t_i \alpha_i$, such that

$$\mu_{\alpha,A} = \sum t_i \mu_{\alpha_i,A}$$

for every observable A.

AXIOM 9. If $\{q_i\}$ is a net of questions, such that the net $\{m_{q_i}\}$ of functions converges pointwise to a function m on S, then there exists a question q such that $m_q = m$.

AXIOM 10. If $\{\alpha_i\}$ is a net of states, for which the net $\{\mu_{\alpha_i}\}$ of characters on Q converges pointwise to a character μ, then there exists a state α such that $\mu_\alpha = \mu$.

AXIOM 11. If μ is a character of Q, then there exists a state α for which $\mu_\alpha = \mu$.

EXERCISE 7.18. Discuss the intuitive legitimacy of Axioms 8, 9, 10, and 11.

AXIOM 12. If p and q are (compatible) questions, such that $p \leq q$ and $p \leq \tilde{q}$, then $p = 0$.

EXERCISE 7.19. (a) Discuss the intuitive legitimacy of Axiom 12.

(b) Suppose that for each question q there exists a state α such that $m_q(\alpha) > 1/2$. Show that Axiom 12 must then be valid.

CHAPTER VIII

HILBERT SPACES

DEFINITION Let X and Y be two complex vector spaces. A map $T : X \to Y$ is called a *conjugate-linear* transformation if it is a real-linear transformation from X into Y, and if

$$T(\lambda x) = \bar{\lambda} T(x)$$

for all $x \in X$ and $\lambda \in \mathbb{C}$.

Let X be a complex vector space. An *inner product* or *Hermitian form* on X is a mapping from $X \times X$ into $\mathbb{C}$ (usually denoted by (x, y)) which satisfies the following conditions:

(1) $(x, y) = \overline{(y, x)}$ for all $x, y \in X$.
(2) For each fixed $y \in X$, the map $x \to (x, y)$ is a linear functional on X.
(3) $(x, x) > 0$ for all nonzero $x \in X$.

Note that conditions 1 and 2 imply that for each fixed vector x the map $y \to (x, y)$ is conjugate-linear. It also follows from condition 1 that $(0, x) = 0$ for all $x \in X$.

The complex vector space X, together with an inner product $(,)$, is called an *inner product space*.

REMARK. We treat here primarily complex inner product spaces and complex Hilbert spaces. Corresponding definitions can be given for real inner product spaces and real Hilbert spaces, and the results about these spaces are occasionally different from the complex cases.

EXERCISE 8.1. (a) Let X be the complex vector space of all continuous complex-valued functions on $[0,1]$, and define

$$(f,g) = \int_0^1 f(x)\overline{g(x)}\,dx.$$

Show that X, with this definition of $(\ ,\ )$, is an inner product space.
  (b) Let $X = \mathbb{C}^n$, and define

$$(x,y) = \sum_{j=1}^n x_j \overline{y_j},$$

where $x = (x_1,\ldots,x_n)$ and $y = (y_1,\ldots,y_n)$. Prove that X, with this definition of $(\ ,\ )$, is an inner product space.
  (c) (General l^2) Let μ be counting measure on a countable set (sequence) S. Let $X = L^2(\mu)$, and for $f,g \in X$ define

$$(f,g) = \int_S f(s)\overline{g(s)}\,d\mu(s) = \sum_{s \in S} f(s)\overline{g(s)}.$$

Prove that X is an inner product space with respect to this definition.
  (d) Specialize the inner product space defined in part c to the two cases first where S is the set of nonnegative integers and then second where S is the set $\mathbb{Z}$ of all integers.

THEOREM 8.1. *Let X be an inner product space.*
  (1) *(Cauchy-Schwarz Inequality) For all $x,y \in X$,*

$$|(x,y)| \leq \sqrt{(x,x)}\sqrt{(y,y)}.$$

  (2) *The assignment $x \to \sqrt{(x,x)}$ is a norm on X, and X equipped with this norm is a normed linear space.*

PROOF. Fix x and y in X. If either x or y is 0, then part 1 is immediate. Otherwise, define a function f of a complex variable λ by

$$f(\lambda) = (x + \lambda y\ ,\ x + \lambda y),$$

and note that $f(\lambda) \geq 0$ for all λ. We have that

$$f(\lambda) = (x,x) + \lambda(y,x) + \overline{\lambda}(x,y) + (y,y)|\lambda|^2.$$

Substituting $\lambda = -(x, y)/(y, y)$, and using the fact that $f(\lambda) \geq 0$ for all λ, the general case of part 1 follows.

To see that $x \to \sqrt{(x, x)}$ defines a norm $\|x\|$ on X, we need only check that $\|x + y\| \leq \|x\| + \|y\|$. But

$$
\begin{aligned}
\|x + y\|^2 &= (x + y \ , \ x + y) \\
&= (x, x) + 2\Re((x, y)) + (y, y) \\
&\leq \|x\|^2 + 2|(x, y)| + \|y\|^2 \\
&\leq \|x\|^2 + 2\|x\|\|y\| + \|y\|^2 \\
&= (\|x\| + \|y\|)^2,
\end{aligned}
$$

which completes the proof of part 2.

EXERCISE 8.2. (a) Show that equality holds in the Cauchy-Schwarz inequality, i.e.,

$$
|(x, y)| = \|x\|\|y\|,
$$

if and only if one of the vectors is a scalar multiple of the other. Conclude that equality holds in the triangle inequality for the norm if and only if one of the vectors is a nonnegative multiple of the other.

(b) Let y and z be elements of an inner product space X. Show that $y = z$ if and only if $(x, y) = (x, z)$ for all $x \in X$.

(c) Prove the *polarization identity* and the *parallelogram law* in an inner product space X; i.e., show that for $x, y \in X$, we have

$$
(x, y) = (1/4) \sum_{j=0}^{3} i^j \|x + i^j y\|^2
$$

and

$$
\|x + y\|^2 + \|x - y\|^2 = 2(\|x\|^2 + \|y\|^2).
$$

(d) Suppose X and Y are inner product spaces and that T is a linear isometry of X into Y. Prove that T preserves inner products. That is, if $x_1, x_2 \in X$, then

$$
(T(x_1), T(x_2)) = (x_1, x_2).
$$

(e) Suppose X is an inner product space, that Y is a normed linear space, and that T is a linear isometry of X onto Y. Show that there exists an inner product $(,)$ on Y such that $\|y\| = \sqrt{(y, y)}$ for every $y \in Y$; i.e., Y is an inner product space and the norm on Y is determined by that inner product.

(f) Suppose Y is a normed linear space whose norm satisfies the parallelogram law:

$$\|x + y\|^2 + \|x - y\|^2 = 2(\|x\|^2 + \|y\|^2)$$

for all $x, y \in Y$. Show that there exists an inner product $(\ ,\ )$ on Y such that $\|y\| = \sqrt{(y, y)}$ for every $y \in Y$; i.e., Y is an inner product space and the given norm on Y is determined by that inner product. HINT: Use the polarization identity to define (x, y). Show directly that $(y, x) = \overline{(x, y)}$ and that $(x, x) > 0$ if $x \neq 0$. For a fixed y, define $f(x) = (x, y)$. To see that f is linear, first use the parallelogram law to show that

$$f(x + x') + f(x - x') = 2f(x),$$

from which it follows that $f(\lambda x) = \lambda f(x)$ for all $x \in Y$ and $\lambda \in \mathbb{C}$. Then, for arbitrary elements $u, v \in Y$, write $u = x + x'$ and $v = x - x'$.

(g) Show that the inner product is a continuous function of $X \times X$ into $\mathbb{C}$. In particular, the map $x \to (x, y)$ is a continuous linear functional on X for every fixed $y \in X$.

DEFINITION. A (complex) *Hilbert space* is an inner product space that is complete in the metric defined by the norm that is determined by the inner product. An inner product space X is called *separable* if there exists a countable dense subset of the normed linear space X.

REMARK. Evidently, a Hilbert space is a special kind of complex Banach space. The inner product spaces and Hilbert spaces we consider will always be assumed to be separable.

EXERCISE 8.3. Let X be an inner product space. Show that any subspace $M \subseteq X$ is an inner product space, with respect to the restriction of the inner product on X, and show that a closed subspace of a Hilbert space is itself a Hilbert space. If M is a closed subspace of a Hilbert space H, is the quotient space H/M necessarily a Hilbert space?

DEFINITION. Let X be an inner product space. Two vectors x and y in X are called *orthogonal* or *perpendicular* if $(x, y) = 0$. Two subsets S and T are *orthogonal* if $(x, y) = 0$ for all $x \in S$ and $y \in T$. If S is a subset of X, then $S^{\perp}$ will denote what we call the *orthogonal complement* to S and consists of the elements $x \in X$ for which $(x, y) = 0$ for all $y \in S$. A collection of pairwise orthogonal unit vectors is called an *orthonormal* set.

EXERCISE 8.4. Let X be an inner product space.

(a) Show that a collection $x_1, \ldots, x_n$ of nonzero pairwise orthogonal vectors in X is a linearly independent set. Verify also that

$$\left\| \sum_{i=1}^n c_i x_i \right\|^2 = \sum_{i=1}^n |c_i|^2 \|x_i\|^2.$$

(b) (Gram-Schmidt Process) Let $x_1, \ldots$ be a (finite or infinite) sequence of linearly independent vectors in X. Show that there exists a sequence $w_1, \ldots$ of orthonormal vectors such that the linear span of $x_1, \ldots, x_i$ coincides with the linear span of $w_1, \ldots, w_i$ for all $i \geq 1$. HINT: Define the w_i's recursively by setting

$$w_i = \frac{x_i - \sum_{k=1}^{i-1}(x_i, w_k)w_k}{\|x_i - \sum_{k=1}^{i-1}(x_i, w_k)w_k\|}.$$

(c) Show that if X is a separable Hilbert space, then there exists an orthonormal sequence $\{x_i\}$ whose linear span is dense in H.

(d) If M is a subspace of X, show that the set $M^\perp$ is a closed subspace of X. Show further that $M \cap M^\perp = \{0\}$.

(e) Let $X = C([01])$ be the inner product space from part a of Exercise 8.1. For each $0 < t < 1$, let M_t be the set of all $f \in X$ for which $\int_0^t f(x)\,dx = 0$. Show that the collection $\{M_t\}$ forms a pairwise distinct family of closed subspaces of X. Show further that $M_t^\perp = \{0\}$ for all $0 < t < 1$. Conclude that, in general, the map $M \to M^\perp$ is not 1-1.

(f) Suppose X is a Hilbert Space. If M and N are orthogonal closed subspaces of X, show that the subspace $M+N$, consisting of the elements $x + y$ for $x \in M$ and $y \in N$, is a closed subspace.

THEOREM 8.2. *Let H be a separable infinite-dimensional (complex) Hilbert space. Then*

(1) *Every orthonormal set must be countable.*

(2) *Every orthonormal set in H is contained in a (countable) maximal orthonormal set. In particular, there exists a (countable) maximal orthonormal set.*

(3) *If $\{\phi_1, \phi_2, \ldots\}$ is an orthonormal sequence in H, and $\{c_1, c_2, \ldots\}$ is a square summable sequence of complex numbers, then the infinite series $\sum c_n \phi_n$ converges to an element in H.*

(4) *(Bessel's Inequality) If $\phi_1, \phi_2, \ldots$ is an orthonormal sequence in H, and if $x \in H$, then*

$$\sum_n |(x, \phi_n)|^2 \leq \|x\|^2.$$

(5) If $\{\phi_n\}$ denotes a maximal orthonormal sequence (set) in H, then every element $x \in H$ is uniquely expressible as a (infinite) sum

$$x = \sum_n c_n \phi_n,$$

where the sequence $\{c_n\}$ is a square summable sequence of complex numbers. Indeed, we have that $c_n = (x, \phi_n)$.

(6) If $\{\phi_n\}$ is any maximal orthonormal sequence in H, and if $x, y \in H$, then

$$(x, y) = \sum_n (x, \phi_n)\overline{(y, \phi_n)}.$$

(7) (Parseval's Equality) For any $x \in H$ and any maximal orthonormal sequence $\{\phi_n\}$, we have

$$\|x\|^2 = \sum_n |(x, \phi_n)|^2.$$

(8) Let $\{\phi_1, \phi_2, \ldots\}$ be a maximal orthonormal sequence in H, and define $T : l^2 \to H$ by

$$T(\{c_n\}) = \sum_{n=1}^{\infty} c_n \phi_n.$$

Then T is an isometric isomorphism of l^2 onto H. Consequently, any two separable infinite-dimensional Hilbert spaces are isometrically isomorphic.

PROOF. Suppose an orthonormal set in H is uncountable. Then, since the distance between any two distinct elements of this set is $\sqrt{2}$, it follows that there exists an uncountable collection of pairwise disjoint open subsets of H, whence H is not separable. Hence, any orthonormal set must be countable, i.e., a sequence.

Let S be an orthonormal set in H. The existence of a maximal orthonormal set containing S now follows from the Hausdorff maximality principle, applied to the collection of all orthonormal sets in H that contain S.

Next, let $\{\phi_1, \phi_2, \ldots\}$ be an orthonormal sequence, and let $x \in H$ be given. For each positive integer i, set $c_i = (x, \phi_i)$. Then, for each positive integer n We have

$$
\begin{aligned}
0 \leq \|x - \sum_{i=1}^{n} c_i \phi_i\|^2 \\
= ((x - \sum_{i=1}^{n} c_i \phi_i), (x - \sum_{j=1}^{n} c_j \phi_j)) \\
= (x, x) - \sum_{j=1}^{n} \overline{c_j}(x, \phi_j) - \sum_{i=1}^{n} c_i(\phi_i, x) + \sum_{i=1}^{n} \sum_{j=1}^{n} c_i \overline{c_j}(\phi_i, \phi_j) \\
= (x, x) - \sum_{j=1}^{n} \overline{c_j} c_j - \sum_{i=1}^{n} c_i \overline{c_i} + \sum_{i=1}^{n} \sum_{j=1}^{n} c_i \overline{c_j} \delta_{ij} \\
= \|x\|^2 - \sum_{i=1}^{n} |c_i|^2.
\end{aligned}
$$

Since this is true for an arbitrary n, Bessel's inequality follows.

If $\{\phi_1, \phi_2, \dots\}$ is an orthonormal sequence and $\{c_1, c_2, \dots\}$ is a square summable sequence of complex numbers, then the sequence of partial sums of the infinite series

$$
\sum_{n=1}^{\infty} c_n \phi_n
$$

is a Cauchy sequence. Indeed,

$$
\|\sum_{n=1}^{j} c_n \phi_n - \sum_{n=1}^{k} c_n \phi_n\|^2 = \sum_{n=k+1}^{j} |c_n|^2.
$$

See part a of Exercise 8.4. This proves part 4.

Now, if $\{\phi_1, \phi_2, \dots\}$ is a maximal orthonormal sequence in H, and x is an element of H, we have from Bessel's inequality that $\sum_n |(x, \phi_n)|^2$ is finite, and therefore $\sum(x, \phi_n)\phi_n$ converges in H by part 4. If we define

$$
y = \sum_n (x, \phi_n)\phi_n.
$$

Clearly $((x - y), \phi_n) = 0$ for all n, implying that, if $x - y \neq 0$, then $(x - y)/\|x - y\|$ is a unit vector that is orthogonal to the set $\{\phi_n\}$. But since this set is maximal, no such vector can exist, and we must have $x = y$ as desired. To see that this representation of x as an infinite series is unique, suppose $x = \sum c'_n \phi_n$, where $\{c'_n\}$ is a square-summable sequence of complex numbers. Then, for each k, we have

$$(x, \phi_k) = \sum_n c'_n (\phi_n, \phi_k) = c'_k,$$

showing the uniqueness of the coefficients.

Because the inner product is continuous in both variables, we have that

$$(x, y) = \sum_n \sum_k ((x, \phi_n)\phi_n, (y, \phi_k)\phi_k) = \sum_n (x, \phi_n)\overline{(y, \phi_n)},$$

proving part 6.

Parseval's equality follows from part 6 by setting $y = x$.

Part 7 is now immediate, and this completes the proof.

DEFINITION. We call a maximal orthonormal sequence in a separable Hilbert space H an *orthonormal basis* of H.

EXERCISE 8.5. (a) Prove that $L^2[0, 1]$ is a Hilbert space with respect to the inner product defined by

$$(f, g) = \int_0^1 f(x)\overline{g(x)}\, dx.$$

(b) For each integer n, define an element $\phi_n \in L^2[0, 1]$ by $\phi_n(x) = e^{2\pi i n x}$. Show that the ϕ_n's form an orthonormal sequence in $L^2[0, 1]$.

(c) For each $0 < r < 1$, define a function k_r on $[0, 1]$ by

$$k_r(x) = \frac{1 - r^2}{1 + r^2 - 2r\cos(2\pi x)}.$$

(See part d of Exercise 6.7.) Show that

$$k_r(x) = \sum_{n=-\infty}^{\infty} r^{|n|}\phi_n(x),$$

whence $\int_0^1 k_r(x)\, dx = 1$ for every $0 < r < 1$. Show further that if $f \in L^2[0, 1]$, then

$$f = \lim_{r \to 1} k_r * f,$$

where the limit is taken in L^2, and where $*$ denotes convolution; i.e.,

$$(k_r * f)(x) = \int_0^1 k_r(x - y)f(y)\, dy.$$

(d) Suppose $f \in L^2[0,1]$ satisfies $(f, \phi_n) = 0$ for all n. Show that f is the 0 element of $L^2[0,1]$. HINT: $(k_r * f)(x) = 0$ for every $r < 1$.

(e) Conclude that the set $\{\phi_n\}$ forms an orthonormal basis for $L^2[0,1]$.

(f) Using $f(x) = x$, show that

$$\sum_{n=1}^{\infty} 1/n^2 = \pi^2/6.$$

Then, using $f(x) = x^2 - x$, show that

$$\sum_{n=1}^{\infty} 1/n^4 = \pi^4/90.$$

HINT: Parseval's equality.

(g) Let M be the set of all functions $f = \sum c_n \phi_n$ in $L^2[0,1]$ for which $c_{2n+1} = 0$ for all n, and let N be the set of all functions $g = \sum c_n \phi_n$ in $L^2[0,1]$ for which $c_{2n} = (1 + |n|)c_{2n+1}$ for all n. Prove that both M and N are closed subspaces of $L^2[0,1]$.

(h) For M and N as in part g, show that $M + N$ contains each ϕ_n and so is a dense subspace of $L^2[0,1]$. Show further that if $h = \sum_n c_n \phi_n \in M + N$, then

$$\sum_n n^2 |c_{2n+1}|^2 < \infty.$$

(i) Conclude that the sum of two arbitrary closed subspaces of a Hilbert space need not be closed. Compare this with part f of Exercise 8.4.

THEOREM 8.3. (Projection Theorem) *Let M be a closed subspace of a separable Hilbert space H. Then:*

(1) *H is the direct sum $H = M \oplus M^\perp$ of the closed subspaces M and $M^\perp$; i.e., every element $x \in H$ can be written uniquely as $x = y + z$ for $y \in M$ and $z \in M^\perp$.*

(2) *For each $x \in H$ there exists a unique element $y \in M$ for which $x - y \in M^\perp$. We denote this unique element y by $p_M(x)$.*

(3) *The assignment $x \to p_M(x)$ of part 2 defines a continuous linear transformation p_M of H onto M that satisfies $p_M^2 = p_M$.*

PROOF. We prove part 1 and leave the rest of the proof to an exercise. Let $\{\phi_n\}$ be a maximal orthonormal sequence in the Hilbert space M, and extend this set, by Theorem 8.2, to a maximal orthonormal sequence $\{\phi_n\} \cup \{\psi_k\}$ in H. If $x \in H$, then, again according to Theorem 8.2, we have

$$x = \sum_n (x, \phi_n)\phi_n + \sum_k (x, \psi_k)\psi_k = y + z,$$

where $y = \sum_n (x, \phi_n)\phi_n$ and $z = \sum_k (x, \psi_k)\psi_k$. Clearly, $y \in M$ and $z \in M^\perp$. If $x = y' + z'$, for $y' \in M$ and $z' \in M^\perp$, then the element $y - y' = z' - z$ belongs to $M \cap M^\perp$, whence is 0. This shows the uniqueness of y and z and completes the proof of part 1.

EXERCISE 8.6. (a) Complete the proof of the preceding theorem.

(b) For p_M as in part 3 of the preceding theorem, show that

$$\|p_M(x)\| \leq \|x\|$$

for all $x \in H$, i.e., p_M is norm-decreasing.

(c) Again, for p_M as in part 3 of the preceding theorem, show that

$$(p_M(x), y) = (x, p_M(y))$$

for all $x, y \in H$.

(d) Let S be a subset of a Hilbert space H. Show that $(S^\perp)^\perp$ is the smallest closed subspace of H that contains S. Conclude that if M is a closed subspace of a Hilbert space H, then $M = (M^\perp)^\perp$.

(e) Let M be a subspace of a Hilbert space H. Show that M is dense in H if and only if $M^\perp = \{0\}$. Give an example of a proper closed subspace M of an inner product space X (necessarily not a Hilbert space) for which $M^\perp = \{0\}$.

(f) Let M be a closed subspace of a Hilbert space H. Define a map $T : M^\perp \to H/M$ by $T(x) = x + M$. Prove that T is an isometric isomorphism of $M^\perp$ onto H/M. Conclude then that the quotient space H/M, known to be a Banach space, is in fact a Hilbert space.

DEFINITION. If M is a closed subspace of a separable Hilbert space H, then the transformation p_M of the preceding theorem is called the *projection* of H onto M.

REMARK. The set $\mathcal{M}$ of all closed subspaces of a Hilbert space H is a candidate for the set Q of questions in our mathematical model of experimental science. (See Chapter VII.) Indeed, $\mathcal{M}$ is obviously a partially-ordered set by inclusion; it contains a maximum element H and a minimum element $\{0\}$; the sum of two orthogonal closed subspaces is a closed subspace, so that there is a notion of summability for certain pairs of elements of $\mathcal{M}$; each element $M \in \mathcal{M}$ has a complement $M^{\perp}$ satisfying $M + M^{\perp} = H$. Also, we may define M and N to be compatible if there exist four pairwise orthogonal closed subspaces $M_1, \ldots, M_4$ satisfying

(1) $M = M_1 + M_2$.
(2) $N = M_2 + M_3$.
(3) $M_1 + M_2 + M_3 + M_4 = H$.

We study this candidate for Q in more detail later by putting it in 1-1 correspondence with the corresponding set of projections.

DEFINITION. Let $\{H_n\}$ be a sequence of Hilbert spaces. By the *Hilbert space direct sum* $\bigoplus H_n$ of the H_n's, we mean the subspace of the direct product $\prod_n H_n$ consisting of the sequences $\{x_n\}$, for which $x_n \in H_n$ for each n, and for which $\sum_n \|x_n\|^2 < \infty$.

EXERCISE 8.7. (a) Prove that the Hilbert space direct sum $\bigoplus H_n$ of Hilbert spaces $\{H_n\}$ is a Hilbert space, where the vector space operations are componentwise and the inner product is defined by

$$(\{x_n\}, \{y_n\}) = \sum_n (x_n, y_n).$$

Show that the ordinary (algebraic) direct sum of the vector spaces $\{H_n\}$ can be naturally identified with a dense subspace of the Hilbert space direct sum of the H_n's. Verify that if each H_n is separable then so is $\bigoplus_n H_n$.

(b) Suppose $\{M_n\}$ is a pairwise orthogonal sequence of closed subspaces of a Hilbert space H and that M is the smallest closed subspace of H that contains each M_n. Construct an isometric isomorphism between M and the Hilbert space direct sum $\bigoplus M_n$, where we regard each closed subspace M_n as a Hilbert space in its own right.

THEOREM 8.4. (Riesz Representation Theorem for Hilbert Space) *Let H be a separable Hilbert space, and let f be a continuous linear functional on H. Then there exists a unique element y_f of H for which $f(x) = (x, y_f)$ for all $x \in H$. That is, the linear functional f can be represented as an inner product. Moreover, the map $f \to y_f$ is a conjugate-linear isometric isomorphism of the conjugate space H^* onto H.*

PROOF. Let $\{\phi_1, \phi_2, \ldots\}$ be a maximal orthonormal sequence in H, and for each n, define $c_n = f(\phi_n)$. Note that $|c_n| \leq \|f\|$ for all n; i.e., the sequence $\{c_n\}$ is bounded. For any positive integer n, write $w_n = \sum_{j=1}^{n} \overline{c_j} \phi_j$, and note that

$$\|w_n\|^2 = \sum_{j=1}^{n} |c_j|^2 = |f(w_n)| \leq \|f\| \|w_n\|,$$

whence

$$\sum_{j=1}^{n} |c_j|^2 \leq \|f\|^2,$$

showing that the sequence $\{\overline{c_n}\}$ belongs to l^2. Therefore, the series $\sum_{n=1}^{\infty} \overline{c_n} \phi_n$ converges in H to an element y_f. We see immediately that $(y_f, \phi_n) = \overline{c_n}$ for every n, and that $\|y_f\| \leq \|f\|$. Further, for each $x \in H$, we have by Theorem 8.2 that

$$\begin{aligned} f(x) &= f\left(\sum (x, \phi_n)\phi_n\right) \\ &= \sum (x, \phi_n)c_n \\ &= \sum (x, \phi_n)\overline{(y_f, \phi_n)} \\ &= (x, y_f), \end{aligned}$$

showing that $f(x) = (x, y_f)$ as desired.

From the Cauchy-Schwarz inequality, we then see that $\|f\| \leq \|y_f\|$, and we have already seen the reverse inequality above. Hence, $\|f\| = \|y_f\|$. We leave the rest of the proof to the exercise that follows.

EXERCISE 8.8. (a) Prove that the map $f \to y_f$ of the preceding theorem is conjugate linear, isometric, and onto H. Conclude that the map $f \to y_f$ is a conjugate-linear, isometric isomorphism of the conjugate space H^* of H onto H. Accordingly, we say that a Hilbert space is *self dual.*

(b) Let H be a Hilbert space. Show that a net $\{x_\alpha\}$ of vectors in H converges to an element x in the weak topology of H if and only if

$$(x, y) = \lim_{\alpha} (x_\alpha, y)$$

for every $y \in H$.

(c) Show that the map $f \rightarrow y_f$ of the preceding theorem is a homeomorphism of the topological vector space (H^*, W^*) onto the topological vector space (H, W).

(d) Let H be a separable Hilbert space. Prove that the closed unit ball in H is compact and metrizable in the weak topology.

(e) Let H be a separable Hilbert space and let $\{x_n\}$ be a sequence of vectors in H. If $\{x_n\}$ converges weakly to an element $x \in H$, show that the sequence $\{x_n\}$ is uniformly bounded in norm. Conversely, if the sequence $\{x_n\}$ is uniformly bounded in norm, prove that there exists a subsequence $\{x_{n_k}\}$ of $\{x_n\}$ that is weakly convergent. HINT: Uniform Boundedness Principle and Alaoglu's Theorem.

DEFINITION. Let H be a Hilbert space, and let $B(H)$ denote the set $L(H, H)$ of all bounded linear transformations of H into itself. If $T \in B(H)$ and $x, y \in H$, we call the number $(T(x), y)$ a matrix coefficient for T.

Let T be an element of $B(H)$. Define, as in Chapter IV, $\|T\|$ by

$$\|T\| = \sup_{\substack{x \in H \\ \|x\| \le 1}} \|T(x)\|.$$

EXERCISE 8.9. (a) For $T \in B(H)$, show that

$$\|T\| = \sup_{\substack{x, y \in H \\ \|x\| \le 1, \|y\| \le 1}} |(T(x), y)|.$$

(b) For $T \in B(H)$ and $x, y \in H$, prove the following polarization identity:

$$(T(x), y) = (1/4) \sum_{j=0}^{3} i^j (T(x + i^j y), (x + i^j y)).$$

(c) If $S, T \in B(H)$, show that $\|TS\| \le \|T\|\|S\|$. Conclude that $B(H)$ is a *Banach algebra*; i.e., $B(H)$ is a Banach space on which there is also defined an associative multiplication $\times$, which is distributive over addition, and which satisfies $\|T \times S\| \le \|T\|\|S\|$.

(d) If $S, T \in B(H)$ satisfy $(T(x), y) = (S(x), y)$ for all $x, y \in H$ (i.e., they have the same set of matrix coefficients), show that $S = T$. Show further that $T = S$ if and only if $(T(x), x) = (S(x), x)$ for all $x \in H$. (This is a result that is valid in complex Hilbert spaces but is not valid in Hilbert spaces over the real field. Consider the linear transformation on $\mathbb{R}^2$ determined by the matrix $\begin{bmatrix} 0 & 1 \\ -1 & 0 \end{bmatrix}$.)

(e) If F, G are continuous linear transformations of H into any topological vector space X, and if $F(\phi_n) = G(\phi_n)$ for all ϕ_n in a maximal orthonormal sequence, show that $F = G$.

THEOREM 8.5. *Let H be a complex Hilbert space, and let L be a mapping of $H \times H$ into $\mathbb{C}$ satisfying:*

(1) *For each fixed y, the map $x \to L(x, y)$ is a linear functional on H.*

(2) *For each fixed x, the map $y \to L(x, y)$ is a conjugate linear transformation of H into $\mathbb{C}$.*

(3) *There exists a positive constant M such that*

$$|L(x, y)| \leq M \|x\| \|y\|$$

for all $x, y \in H$.

(Such an L is called a bounded Hermitian form on H.) Then there exists a unique element $S \in B(H)$ such that

$$L(x, y) = (x, S(y))$$

for all $x, y \in H$.

PROOF. For each fixed $y \in H$, we have from assumptions (1) and (3) that the map $x \to L(x, y)$ is a continuous linear functional on H. Then, by the Riesz representation theorem (Theorem 8.4), there exists a unique element $z \in H$ for which $L(x, y) = (x, z)$ for all $x \in H$. We denote z by $S(y)$, and we need to show that S is a continuous linear transformation of H into itself.

Clearly,

$$(x, S(y_1 + y_2)) = L(x, y_1 + y_2)$$
$$= L(x, y_1) + L(x, y_2)$$
$$= (x, S(y_1)) + (x, S(y_2))$$
$$= (x, S(y_1) + S(y_2))$$

for all x, showing that $S(y_1 + y_2) = S(y_1) + S(y_2)$. Also,

$$(x, S(\lambda y)) = L(x, \lambda y)$$
$$= \overline{\lambda} L(x, y)$$
$$= \overline{\lambda}(x, S(y))$$
$$= (x, \lambda S(y))$$

for all x, showing that $S(\lambda y) = \lambda S(y)$, whence S is linear.

Now, since $|(x, S(y))| = |L(x, y)| \leq M\|x\|\|y\|$, it follows by setting $x = S(y)$ that S is a bounded operator of norm $\leq M$ on H, as desired.

Finally, the uniqueness of S is evident since any two such operators S_1 and S_2 would have identical matrix coefficients and so would be equal.

DEFINITION. Let T be a bounded operator on a (complex) Hilbert space H. Define a map L_T on $H \times H$ by

$$L_T(x, y) = (T(x), y).$$

By the *adjoint* of T, we mean the unique bounded operator $S = T^*$, whose existence is guaranteed by the previous theorem, that satisfies

$$(x, T^*(y)) = L_T(x, y) = (T(x), y)$$

for all $x, y \in H$.

THEOREM 8.6. *The adjoint mapping $T \to T^*$ on $B(H)$ satisfies the following for all $T, S \in B(H)$ and $\lambda \in \mathbb{C}$.:*

(1) $(T + S)^* = T^* + S^*$.
(2) $(\lambda T)^* = \bar{\lambda} T^*$.
(3) $(TS)^* = S^* T^*$.
(4) $(T^*)^* = T$.
(5) $\|T^*\| = \|T\|$.
(6) $\|T^* T\| = \|T T^*\| = \|T\|^2$.

PROOF. We prove parts 3 and 6 and leave the remaining parts to an exercise.

We have

$$\begin{aligned}
(x, (TS)^*(y)) &= (T(S(x)), y) \\
&= (S(x), T^*(y)) \\
&= (x, S^*(T^*(y))),
\end{aligned}$$

showing part 3.

Next, we have that $\|T^* T\| \leq \|T^*\|\|T\| = \|T\|^2$ by part 5, so to obtain part 6 we need only show the reverse inequality. Thus,

$$\begin{aligned}
\|T\|^2 &= \sup_{\substack{x \in H \\ \|x\| \leq 1}} \|T(x)\|^2 \\
&= \sup_{\substack{x \in H \\ \|x\| \leq 1}} (T(x), T(x)) \\
&= \sup_{\substack{x \in H \\ \|x\| \leq 1}} (x, T^*(T(x))) \\
&\leq \|T^* T\|,
\end{aligned}$$

as desired.

EXERCISE 8.10. Prove the remaining parts of Theorem 8.6.

DEFINITION. Let H be a (complex) Hilbert space. An element $T \in B(H)$ is called *unitary* if it is an isometry of H onto H. A linear transformation U from one Hilbert space H_1 into another Hilbert space H_2 is called a *unitary* map if it is an isometry of H_1 onto H_2.

An element $T \in B(H)$ is called *selfadjoint* or *Hermitian* if $T^* = T$.

An element $T \in B(H)$ is called *normal* if T and T^* commute, i.e., if $TT^* = T^*T$.

An element T in $B(H)$ is called *positive* if $(T(x), x) \geq 0$ for all $x \in H$.

An element $T \in B(H)$ is called *idempotent* if $T^2 = T$.

If $p \in B(H)$ is selfadjoint and idempotent, we say that p is an *orthogonal projection* or (simply) a *projection*.

An *eigenvector* for an operator $T \in B(H)$ is a nonzero vector $x \in H$ for which there exists a scalar λ satisfying $T(x) = \lambda x$. The scalar λ is called an *eigenvalue* for T, and the eigenvector x is said to *belong* to the eigenvalue λ.

EXERCISE 8.11. (a) Prove that the L^2 Fourier transform U is a unitary operator on $L^2(\mathbb{R})$.

(b) Suppose μ and ν are σ-finite measures on a σ-algebra $\mathcal{B}$ of subsets of a set S, and assume that ν is absolutely continuous with respect to μ. Let f denote the Radon-Nikodym derivative of ν with respect to μ, and define $U : L^2(\nu) \to L^2(\mu)$ by

$$U(g) = \sqrt{f}g.$$

Prove that U is a norm-preserving linear transformation of $L^2(\nu)$ into $L^2(\mu)$, and that it is a unitary transformation between these two Hilbert spaces if and only if μ and ν are mutually absolutely continuous.

(c) (Characterization of unitary transformations) Let U be a linear transformation of a Hilbert space H_1 into a Hilbert space H_2. Prove that U is a unitary operator if and only if it is onto H_2 and is *inner-product preserving*; i.e.,

$$(U(x), U(y)) = (x, y)$$

for all $x, y \in H_1$.

(d) (Another characterization of unitary operators) Let U be an element of $B(H)$. Prove that U is a unitary operator if and only if

$$UU^* = U^*U = I.$$

(e) (The bilateral shift) Let $\mathbb{Z}$ denote the set of all integers, let μ be counting measure on $\mathbb{Z}$, and let H be $L^2(\mu)$. Define a transformation U on H by

$$[U(x)]_n = x_{n+1}.$$

Prove that U is a unitary operator on H. Compute its adjoint (inverse) U^*.

(f) (The unilateral shift) Let S be the set of all nonnegative integers, let μ be counting measure on S, and let $H = L^2(\mu)$. Define a transformation T on H by

$$[t(x)]_n = x_{n+1}.$$

Show that T is not a unitary operator. Compute its adjoint T^*.

THEOREM 8.7. Let H be a (complex) Hilbert space.

(1) If $T \in B(H)$, then there exist unique selfadjoint operators T_1 and T_2 such that $T = T_1 + iT_2$. T_1 and T_2 are called respectively the real and imaginary parts of the operator T.

(2) The set of all selfadjoint operators forms a real Banach space with respect to the operator norm, and the set of all unitary operators forms a group under multiplication.

(3) An element $T \in B(H)$ is selfadjoint if and only if $(T(x), x) = (x, T(x))$ for all x in a dense subset of H.

(4) An element $T \in B(H)$ is selfadjoint if and only if $(T(x), x)$ is real for every $x \in H$. If λ is an eigenvalue for a selfadjoint operator T, then λ is real.

(5) Every positive operator is selfadjoint.

(6) Every orthogonal projection is positive.

(7) If T is selfadjoint, then $I \pm iT$ is 1-1, onto, and $\|(I \pm iT)(x)\| \geq \|x\|$ for all $x \in H$, whence $(I \pm iT)^{-1}$ is a bounded operator on H.

(8) If T is selfadjoint, then $u = (I - iT)(I + iT)^{-1}$ is a unitary operator, for which -1 is not an eigenvalue; i.e., $I + u$ is 1-1. Moreover,

$$T = -i(I - u)(I + u)^{-1}.$$

This unitary operator U is called the Cayley transform of T.

(9) A continuous linear transformation $U : H_1 \to H_2$ is unitary if and only if its range is a dense subspace of H_2, and

$$(U(x), U(x)) = (x, x)$$

for all x in a dense subset of H_1.

PROOF. Defining $T_1 = (1/2)(T + T^*)$ and $T_2 = (1/2i)(T - T^*)$, we have that $T = T_1 + iT_2$, and both T_1 and T_2 are selfadjoint. Further, if $T = S_1 + iS_2$, where both S_1 and S_2 are selfadjoint, then $T^* = S_1 - iS_2$, whence $2S_1 = T + T^*$ and $2iS_2 = T - T^*$, from which part 1 follows.

Parts 2 through 6 are left to the exercises.

To see part 7, notice first that

$$
\begin{aligned}
\|(I + iT)(x)\|^2 &= ((I + iT)(x), (I + iT)(x)) \\
&= (x, x) + i(T(x), x) - i(x, T(x)) + (T(x), T(x)) \\
&= \|x\|^2 + \|T(x)\|^2 \\
&\geq \|x\|^2,
\end{aligned}
$$

which implies that $I + iT$ is 1-1 and norm-increasing. Moreover, it follows that the range of $I + iT$ is closed in H. For, if $y \in H$ and $y = \lim y_n = \lim(I + iT)(x_n)$, then the sequence $\{y_n\}$ is Cauchy, and hence the sequence $\{x_n\}$ must be Cauchy by the above inequality. Therefore $\{x_n\}$ converges to an $x \in H$. Then $y = (I + iT)(x)$, showing that the range of $I + iT$ is closed.

If $z \in H$ is orthogonal to the range of $I + iT$, then $((I + iT)(z), z) = 0$, which implies that $(z, z) = -i(T(z), z)$, which can only happen if $z = 0$, since $(T(z), z)$ is real if T is selfadjoint. Hence, the range of $I + iT$ only has 0 in its orthogonal complement; i.e., this range is dense. Since it is also closed, we have that the range of $I + iT = H$, and $I + iT$ is onto. Since $I + iT$ is norm-increasing, we see that $(I + iT)^{-1}$ exists and is norm-decreasing, hence is continuous.

Of course, an analogous argument proves that $(I - iT)^{-1}$ is continuous.

Starting with
$$
(I + iT)(I + iT)^{-1} = I,
$$

we see by taking the adjoint of both sides that

$$
((I + iT)^{-1})^* = (I - iT)^{-1}.
$$

It follows also then that $I - iT$ and $(I + iT)^{-1}$ commute. But now

$$
\begin{aligned}
I &= (I - iT)(I - iT)^{-1}(I + iT)^{-1}(I + iT) \\
&= (I - iT)(I + iT)^{-1}(I - iT)^{-1}(I + iT) \\
&= (I - iT)(I + iT)^{-1}[(I - iT)(I + iT)^{-1}]^*,
\end{aligned}
$$

showing that $u = (I - iT)(I + iT)^{-1}$ is unitary. See part d of Exercise 8.11. Also,

$$I + u = (I + iT)(I + iT)^{-1} + (I - iT)(I + iT)^{-1}$$
$$= 2(I + iT)^{-1},$$

showing that $I + u$ is 1-1. Finally,

$$I - u = (I + iT)(I + iT)^{-1} - (I - iT)(I + iT)^{-1} = 2iT(I + iT)^{-1},$$

whence

$$-i(I - u)(I + u)^{-1} = -i \times 2iT(I + iT)^{-1}(1/2)(I + iT) = T,$$

as desired.

Finally, if a continuous linear transformation $U : H_1 \to H_2$ is onto a dense subspace of H_2, and $(U(x), U(x)) = (x, x)$ for all x in a dense subset of H_1, we have that U is an isometry on this dense subset, whence is an isometry of all of H_1 into H_2. Since H_1 is a complete metric space, it follows that the range of U is complete, whence is a closed subset of H_2. Since this range is assumed to be dense, it follows that the range of $U = H_2$, and U is unitary.

EXERCISE 8.12. Prove parts 2 through 6 of the preceding theorem.

EXERCISE 8.13. Let H be a Hilbert space.
(a) If $\phi(z) = \sum_{n=0}^{\infty} a_n z^n$ is a power series function with radius of convergence r, and if T is an element of $B(H)$ for which $\|T\| < r$, show that the infinite series $\sum_{n=0}^{\infty} a_n T^n$ converges to an element of $B(H)$. (We may call this element $\phi(T)$.)
(b) Use part a to show that $I + T$ has an inverse in $B(H)$ if $\|T\| < 1$.
(c) For each $T \in B(H)$, define

$$e^T = \sum_{n=0}^{\infty} T^n / n!.$$

Prove that

$$e^{T+S} = e^T e^S$$

if T and S commute. HINT: Show that the double series $\sum T^n / n! \times \sum S^j / j!$ converges independent of the arrangement of the terms. Then, rearrange the terms into groups where $n + j = k$.

(d) Suppose T is selfadjoint. Show that

$$e^{iT} = \sum_{n=0}^{\infty} (iT)^n / n!$$

is unitary.

EXERCISE 8.14. (Multiplication Operators) Let (S, μ) be a σ-finite measure space. For each $f \in L^\infty(\mu)$, define the operator m_f on the Hilbert space $L^2(\mu)$ by

$$m_f(g) = fg.$$

These operators m_f are called *multiplication operators*.

(a) Show that each operator m_f is bounded and that

$$\|m_f\| = \|f\|_\infty.$$

(b) Show that $(m_f)^* = m_{\bar{f}}$. Conclude that each m_f is normal, and that m_f is selfadjoint if and only if f is real-valued a.e.μ.

(c) Show that m_f is unitary if and only if $|f| = 1$ a.e.μ.

(d) Show that m_f is a positive operator if and only if $f(x) \geq 0$ a.e.μ.

(e) Show that m_f is a projection if and only if $f^2 = f$ a.e.μ, i.e., if and only if f is the characteristic function of some set E.

(f) Show that λ is an eigenvalue for m_f if and only if $\mu(f^{-1}(\{\lambda\})) > 0$.

(g) Suppose $\phi(z) = \sum_{n=0}^{\infty} a_n z^n$ is a power series function with radius of convergence r, and suppose $f \in L^\infty(\mu)$ satisfies $\|f\|_\infty < r$. Show that $\phi(m_f) = m_{\phi \circ f}$. (See the previous exercise.)

EXERCISE 8.15. Let H be the complex Hilbert space $L^2(\mathbb{R})$. For $f \in L^1(\mathbb{R})$, write T_f for the operator on H determined by convolution by f. That is, for $g \in L^2(\mathbb{R})$, we have $T_f(g) = f * g$. See Theorem 6.2.

(a) Prove that $T_f \in B(H)$ and that the map $f \to T_f$ is a norm-decreasing linear transformation of $L^1(\mathbb{R})$ into $B(H)$. See Theorem 6.2.

(b) For $g, h \in L^2(\mathbb{R})$ and $f \in L^1(\mathbb{R})$, show that

$$(T_f(g), h) = (\hat{f} U(g), U(h)),$$

where $\hat{f}$ denotes the Fourier transform of f and $U(g)$ and $U(h)$ denote the L^2 Fourier transforms of g and h. Conclude that the map $f \to T_f$ is 1-1.

(c) Show that $T_f^* = T_{f^*}$, where $f^*(x) = \overline{f(-x)}$.

(d) Show that
$$T_{f_1 * f_2} = T_{f_1} \circ T_{f_2}$$
for all $f_1, f_2 \in L^1(\mathbb{R})$. Conclude that T_f is always a normal operator, and that it is selfadjoint if and only if $f(-x) = \overline{f(x)}$ for almost all x. HINT: Fubini's Theorem.

(e) Prove that T_f is a positive operator if and only if $\hat{f}(\xi) \geq 0$ for all $\xi \in \mathbb{R}$.

(f) Show that T_f is never a unitary operator and is never a nonzero projection. Can T_f have any eigenvectors?

We return now to our study of the set $\mathcal{M}$ of all closed subspaces of a Hilbert space H. The next theorem shows that $\mathcal{M}$ is in 1-1 correspondence with a different, and perhaps more tractable, set $\mathcal{P}$.

THEOREM 8.8. Let p be an orthogonal projection on a Hilbert space H, let M_p denote the range of p and let K_p denote the kernel of p. Then:

(1) $x \in M_p$ if and only if $x = p(x)$.
(2) $M_p = K_p^{\perp}$, whence M_p is a closed subspace of H. Moreover, p is the projection of H onto M_p.
(3) The assignment $p \to M_p$ is a 1-1 correspondence between the set $\mathcal{P}$ of all orthogonal projections on H and the set $\mathcal{M}$ of all closed subspaces of H.
(4) M_p and M_q are orthogonal subspaces if and only if $pq = qp = 0$ which implies that $p + q$ is a projection. In fact, $M_{p+q} = M_p + M_q$.
(5) $M_p \subseteq M_q$ if and only if $pq = qp = p$, which implies that $r = q - p$ is a projection, and $q = p + r$.

PROOF. We leave the proof of part 1 to the exercise that follows. If $x \in M_p$, and $y \in K_p$, then, $x = p(x)$ by part 1. Therefore,
$$(x, y) = (p(x), y) = (x, p(y)) = 0,$$
showing that $M_p \subseteq K_p^{\perp}$. Conversely, if $x \in K_p^{\perp}$, then $x - p(x)$ is also in $K_p^{\perp}$. But
$$p(z - p(z)) = p(z) - p^2(z) = 0$$
for any $z \in H$, whence $x - p(x) \in K_p \cap K_p^{\perp}$, and this implies that $x = p(x)$, and $x \in M_p$. This proves the first part of 2. We see also that for any $z \in H$ we have that
$$z = p(z) + (z - p(z)),$$

and that $p(z) \in M_p$, and $z - p(z) \in K_p$. It follows then that p is the projection of H onto the closed subspace M_p. See the Projection Theorem (8.3).

Part 3 follows directly from Theorem 8.3.

Let M_p and M_q be orthogonal subspaces. If x is any element of H, then $q(x) \in M_q$ and $M_q \subseteq K_p$. Therefore $p(q(x)) = 0$ for every $x \in H$; i.e., $pq = 0$. A similar calculation shows that $qp = 0$. Then it follows directly that $p + q$ is selfadjoint and that $(p + q)^2 = p + q$. Conversely, if $pq = qp = 0$, then $M_p \subseteq K_q$, whence M_p is orthogonal to M_q.

We leave part 5 to the exercises.

EXERCISE 8.16. (a) Prove parts 1 and 5 of the preceding theorem.

(b) Let p be a projection with range M_p. Show that a vector x belongs to M_p if and only if $\|p(x)\| = \|x\|$.

REMARK. We now examine the set $\mathcal{P}$ of all projections on a separable complex Hilbert space H as a candidate for the set Q of all questions in our development of axiomatic experimental science. The preceding theorem shows that $\mathcal{P}$ is in 1-1 correspondence with the set $\mathcal{M}$ of all closed subspaces, and we saw earlier that $\mathcal{M}$ could serve as a model for Q. The following theorem spells out the properties of $\mathcal{P}$ that are relevant if we wish to use $\mathcal{P}$ as a model for Q.

THEOREM 8.9. *Consider the set $\mathcal{P}$ of all projections on a separable complex Hilbert space H as being in 1-1 correspondence with the set $\mathcal{M}$ of all closed subspaces of H, and equip $\mathcal{P}$ with the notions of partial order, complement, orthogonality, sum, and compatibility coming from this identification with $\mathcal{M}$. Then:*

(1) $p \le q$ *if and only if* $pq = qp = p$.

(2) p *and* q *are orthogonal if and only if* $pq = qp = 0$.

(3) p *and* q *are summable if and only if they are orthogonal.*

(4) p *and* q *are compatible if and only if they commute, i.e., if and only if* $pq = qp$.

(5) *If* $\{p_i\}$ *is a sequence of pairwise orthogonal projections, then there exists a (unique) projection* p *such that* $p(x) = \sum_i p_i(x)$ *for all* $x \in H$.

PROOF. Parts 1 and 2 follow from the preceding theorem. It also follows from that theorem that if p and q are orthogonal then $p + q$ is a projection, implying that p and q are summable. Conversely, if p and q are summable, then $p + q$ is a projection, and

$$p + q = (p + q)^2 = p^2 + pq + qp + q^2 = p + q + pq + qp,$$

whence $pq = -qp$. But then

$$-pq = -p^2 q^2$$
$$= -ppqq$$
$$= p(-pq)q$$
$$= pqpq$$
$$= (-qp)(-qp)$$
$$= qpqp$$
$$= q(-qp)p$$
$$= -qp,$$

implying that $pq = qp$. But then $pq = qp = 0$, whence p and q are orthogonal. This completes the proof of part 3.

Suppose now that p and q commute, and write $r_2 = pq$. Let $r_1 = p-r_2$, $r_3 = q - r_2$, and $r_4 = I - r_1 - r_2 - r_3$. It follows directly that the r_i's are pairwise orthogonal projections, that $p = r_1 + r_2$ and that $q = r_2 + r_3$. Hence p and q are compatible. Conversely, if p and q are compatible, and $p = r_1 + r_2$ and $q = r_2 + r_3$, where r_1, r_2, r_3 are pairwise orthogonal projections, then $pq = qp = r_2$ and p and q commute.

Finally, to see part 5, let M be the Hilbert space direct sum $\bigoplus M_{p_i}$ of the closed subspaces $\{M_{p_i}\}$, and let p be the projection of H onto M. Then, if $x \in M^\perp$, we have that $p(x) = p_i(x) = 0$ for all i, whence

$$p(x) = \sum p_i(x).$$

On the other hand, if $x' \in M$, then $x' = \sum x_i'$, where for each i, $x_i' \in M_{p_i}$. Obviously then $p(x') = \sum p_i(x_i') = \sum p_i(x')$. Finally, if $z \in H$, then $z = x + x'$, where $x \in M^\perp$ and $x' \in M$. Clearly, we have $p(z) = \sum p_i(z)$, and the proof is complete.

DEFINITION. Let H be a separable Hilbert space. If $\{p_i\}$ is a sequence of pairwise orthogonal projections in $B(H)$, then the projection $p = \sum_i p_i$ from part 5 of the preceding theorem is called the *sum* of the p_i's.

EXERCISE 8.17. (a) Show that the set $\mathcal{P}$ satisfies all the requirements of the set Q of all questions. (See Chapter VII.)

(b) Show that in $\mathcal{P}$ a stronger property holds than is required for Q. That is, show that a sequence $\{p_i\}$ is mutually summable if and only if it is pairwise orthogonal.

CHAPTER IX

PROJECTION-VALUED MEASURES

DEFINITION Let S be a set and let $\mathcal{B}$ be a σ-algebra of subsets of S. We refer to the elements of $\mathcal{B}$ as *Borel* subsets of S and we call the pair $(S, \mathcal{B})$ a *Borel space*.

If H is a separable (complex) Hilbert space, we say that a mapping $E \to p_E$, of $\mathcal{B}$ into the set $\mathcal{P}$ of projections on H, is a *projection-valued measure* (or an *H-projection-valued measure*) on $(S, \mathcal{B})$ if:

(1) $p_S = I$, and $p_\emptyset = 0$.

(2) If $\{E_i\}$ is a countable collection of pairwise disjoint elements of $\mathcal{B}$, then $\{p_{E_i}\}$ is a pairwise orthogonal collection of projections, and $p_{\cup E_i} = \sum p_{E_i}$.

If p $(E \to p_E)$ is an H-projection-valued measure and M is a closed subspace of H, for which $p_E(M) \subseteq M$ for all $E \in \mathcal{B}$, then M is called an *invariant subspace* for p or simply a *p-invariant* subspace. The assignment $E \to p_E|_M$ is called the *restriction* of p to M. See Exercise 9.1.

Two functions f and g on S are said to *agree a.e.p* if the set E of all x for which $f(x) \neq g(x)$ satisfies $p_E = 0$.

A function $f : S \to \mathbb{C}$ is called a *Borel function* or *$\mathcal{B}$-measurable* if $f^{-1}(U) \in \mathcal{B}$ whenever U is an open subset of $\mathbb{C}$. A complex-valued $\mathcal{B}$-measurable function f is said to belong to $L^\infty(p)$ if there exists a positive real number M such that

$$p_{|f|^{-1}(M,\infty)} = p_{\{x : |f(x)| > M\}} = 0,$$

and the L^∞ *norm* (really only a seminorm) of such a function f is defined to be the infimum of all such numbers M. By $L^\infty(p)$, we mean the vector space (or algebra) of all L^∞ functions f equipped with the ∞-norm. See Exercise 9.1.

If H and H' are two separable Hilbert spaces, and if $E \to p_E$ is an H-projection-valued measure and $E \to p'_E$ is an H'-projection-valued measure, both defined on the same Borel space $(S, \mathcal{B})$, we say that p and p' are *unitarily equivalent* if there exists a unitary transformation $U : H \to H'$ such that

$$U \circ p_E \circ U^{-1} = p'_E$$

for every $E \in \mathcal{B}$.

If we are thinking of the set $\mathcal{P}$ as a model for the set Q of all questions (see Chapter VII), and the Borel space S is the real line $\mathbb{R}$, then the set of projection-valued measures will correspond to the set O of all observables.

EXERCISE 9.1. Let $E \to p_E$ be a projection-valued measure on $(S, \mathcal{B})$.

(a) If $E \in \mathcal{B}$, show that $p_{\tilde{E}} = I - p_E$.

(b) If $E, F \in \mathcal{B}$, show that $p_{E \cap F} = p_E p_F$. HINT: Show first that if $E \cap F = \emptyset$, then p_E and p_F are orthogonal, i.e., that $p_E p_F = p_F p_E = 0$.

(c) If S is the increasing union $\cup E_n$ of elements of $\mathcal{B}$, show that the union of the ranges of the projections p_{E_n} is dense in H. HINT: Write $F_1 = E_1$, and for $n > 1$ define $F_n = E_n - E_{n-1}$. Note that $S = \cup F_n$, whence $x = \sum p_{F_n}(x)$ for each $x \in H$.

(d) Suppose $\{E_n\}$ is a sequence of elements of $\mathcal{B}$ for which $p_{E_n} = 0$ for all n. Prove that $p_{\cup E_n} = 0$.

(e) Show that $\|f\|_\infty$ is a seminorm on $L^\infty(p)$. Show further that $\|fg\|_\infty \le \|f\|_\infty \|g\|_\infty$ for all $f, g \in L^\infty(p)$. If M denotes the subset of $L^\infty(p)$ consisting of the functions f for which $\|f\|_\infty = 0$, i.e., the functions that are 0 a.e.p, prove that $L^\infty(p)/M$ is a Banach space (even a Banach algebra). See part c of Exercise 4.3. Sometimes the notation $L^\infty(p)$ stands for this Banach space $L^\infty(p)/M$.

(f) Suppose M is a closed invariant subspace for p. Show that the assignment $E \to p_E|_M$ is an M-projection-valued measure.

(g) Let $\{H_i\}$ be a sequence of separable Hilbert spaces, and for each i let $E \to p^i_E$ be an H_i-projection-valued measure on the Borel space $(S, \mathcal{B})$. Let $H = \bigoplus H_i$ and define a map $E \to p_E$ of $\mathcal{B}$ into the set of projections on H by

$$p_E = \sum_i p_E^i.$$

Prove that $E \to p_E$ is a projection-valued measure. This projection-valued measure is called the *direct sum* of the projection-valued measures $\{p^i\}$.

THEOREM 9.1. Let $(S, \mathcal{B})$ be a Borel space, let H be a separable Hilbert space, and let $E \to p_E$ be an H-projection-valued measure on $(S, \mathcal{B})$. If $x \in H$, define μ_x on $\mathcal{B}$ by

$$\mu_x(E) = (p_E(x), x).$$

Then μ_x is a finite positive measure on the σ-algebra $\mathcal{B}$ and $\mu_x(S) = \|x\|^2$.

EXERCISE 9.2. (a) Prove Theorem 9.1.

(b) Show that each measure μ_x, as defined in the preceding theorem, is absolutely continuous with respect to p. That is, show that if $p_E = 0$ then $\mu_x(E) = 0$.

(c) Let $S, \mathcal{B}, H$ and p be as in the preceding theorem. If x and y are vectors in H, and if $\mu_{x,y}$ is defined on $\mathcal{B}$ by

$$\mu_{x,y}(E) = (p_E(x), y),$$

show that $\mu_{x,y}$ is a finite complex measure on $\mathcal{B}$. Show also that

$$\|\mu_{x,y}\| \le \|x\|\|y\|.$$

See Exercise 5.12.

(d) Let $S, \mathcal{B}, H, p$, and μ_x be as in the preceding theorem. Suppose p' is any H-projection-valued measure on $\mathcal{B}$ for which $\mu_x(E) = (p_E'(x), x)$ for all $x \in H$. Show that $p' = p$. That is, the measures $\{\mu_x\}$ uniquely determine the projection-valued measure p.

(e) Let ϕ be a $\mathcal{B}$-measurable simple function on S, and suppose

$$\phi = \sum_{i=1}^n a_i \chi_{E_i}$$

and

$$\phi = \sum_{j=1}^m b_j \chi_{F_j}$$

are two different representations of ϕ as finite linear combinations of characteristic functions of elements of $\mathcal{B}$. Prove that for each $x \in H$, we have

$$\sum_{i=1}^{n} a_i p_{E_i}(x) = \sum_{j=1}^{m} b_j p_{F_j}(x).$$

HINT: Show this by taking inner products and using part d.

THEOREM 9.2. Let (S, μ) be a σ-finite measure space, let $\mathcal{B}$ be the σ-algebra of μ-measurable subsets of S, and let $H = L^2(\mu)$. For each measurable set $E \subseteq S$, define p_E to be the projection in $B(H)$ given by $p_E = m_{\chi_E}$. That is,

$$p_E(f) = \chi_E f.$$

Then $E \to p_E$ is a projection-valued measure on H.

DEFINITION. The projection-valued measure of the preceding theorem is called the *canonical projection-valued measure* on $L^2(\mu)$.

EXERCISE 9.3. (a) Prove Theorem 9.2.

(b) Let U denote the L^2 Fourier transform on $L^2(\mathbb{R})$, and, for each Borel subset $E \subseteq \mathbb{R}$, define an operator p_E on $L^2(\mathbb{R})$ by

$$p_E(f) = U^{-1}(\chi_E U(f)).$$

Show that each operator p_E is a projection on $L^2(\mathbb{R})$ and that $E \to p_E$ is a projection-valued measure. Note that this projection-valued measure is unitarily equivalent to the canonical one on $L^2(\mathbb{R})$. Show that $p_{[-1,1]}$ can be expressed as a convolution operator:

$$p_{[-1,1]}f(t) = \int_{-\infty}^{\infty} k(t-s)f(s)\,ds,$$

where k is a certain L^2 function.

(c) Let $(S, \mathcal{B})$ and $(S', \mathcal{B}')$ be two Borel spaces, and let h be a map of S into S' for which $h^{-1}(E') \in \mathcal{B}$ whenever $E' \in \mathcal{B}'$. Such a map h is called a *Borel map* of S into S'. Suppose $E \to p_E$ is an H-projection-valued measure on $(S, \mathcal{B})$, and define a map $E' \to q_{E'}$ on $\mathcal{B}'$ by

$$q_{E'} = p_{h^{-1}(E')}.$$

Prove that $E' \to q_{E'}$ is an H-projection-valued measure on $(S', \mathcal{B}')$. This projection-valued measure q is frequently denoted by $h_*(p)$.

EXERCISE 9.4. Let (S, μ) be a σ-finite measure space, and let $E \rightarrow p_E$ be the canonical projection-valued measure on $L^2(\mu)$. Prove that there exists a vector f in $L^2(\mu)$ such that the linear span of the vectors $p_E(f)$, for E running over the μ-measurable subsets of S, is dense in $L^2(\mu)$. HINT: Do this first for a finite measure μ.

DEFINITION. Let $(S, \mathcal{B})$ be a Borel space, let H be a separable Hilbert space, and let $E \rightarrow p_E$ be an H-projection-valued measure on $(S, \mathcal{B})$. A vector $x \in H$ is called a *cyclic vector* for p if the linear span of the vectors $p_E(x)$, for $E \in \mathcal{B}$, is dense in H.

A vector x is a *separating vector* for p if: $p_E = 0$ if and only if $p_E(x) = 0$.

A vector x is a *supporting vector* for p if the measure μ_x of Theorem 9.1 satisfies: $\mu_x(E) = 0$ if and only if $p_E = 0$.

EXERCISE 9.5. (a) Show that a canonical projection-valued measure has a cyclic vector. (See Exercise 9.4.)

(b) Show that every cyclic vector for a projection-valued measure is a separating vector.

(c) Show that a vector x is a separating vector for a projection-valued measure if and only if it is a supporting vector.

(d) Give an example to show that not every separating vector need be cyclic. HINT: Use a one-point set S and a 2 dimensional Hilbert space.

THEOREM 9.3. *An H-projection-valued measure $E \rightarrow p_E$ on a Borel space $(S, \mathcal{B})$ has a cyclic vector if and only if there exists a finite measure μ on $(S, \mathcal{B})$ such that p is unitarily equivalent to the canonical projection-valued measure on $L^2(\mu)$.*

PROOF. The "if" part follows from part a of Exercise 9.5. Conversely, let x be a cyclic vector for p and write μ for the (finite) measure μ_x of Theorem 9.1 on $\mathcal{B}$. For each $\mathcal{B}$-measurable simple function $\phi = \sum a_i \chi_{E_i}$ on S, define $U(\phi) \in H$ by

$$U(\phi) = \sum a_i p_{E_i}(x).$$

Then $U(\phi)$ is well-defined by part e of Exercise 9.2, and the range of U is dense in H because x is a cyclic vector. It follows directly that U is a well-defined linear transformation of the complex vector space X of all simple $\mathcal{B}$-measurable functions on S into H. Furthermore, writing

$\phi = \sum a_i \chi_{E_i}$, where $E_i \cap E_j = \emptyset$ for $i \neq j$, then

$$\|U(\phi)\|^2 = \left(\sum_i a_i p_{E_i}(x), \sum_j a_j p_{E_j}(x)\right)$$

$$= \sum\sum a_i \overline{a_j} (p_{E_i}(x), p_{E_j}(x))$$

$$= \sum\sum a_i \overline{a_j} (p_{E_j \cap E_i}(x), x)$$

$$= \sum |a_i|^2 (p_{E_i}(x), x)$$

$$= \sum |a_i|^2 \mu(E_i)$$

$$= \int |\phi|^2 \, d\mu$$

$$= \|\phi\|_2^2,$$

showing that U is an isometry of X onto a dense subspace of H.

Therefore, U has a unique extension from the dense subspace X to a unitary operator from all of $L^2(\mu)$ onto all of H.

Finally, if p' denotes the canonical projection-valued measure on $L^2(\mu)$, $\phi = \sum a_i \chi_{E_i}$ is an element of X, and $y = U(\phi)$ is the corresponding element in the range of U on X, we have

$$(U \circ p'_E \circ U^{-1})(y) = (U \circ p'_E)(\phi)$$

$$= U(\chi_E \phi)$$

$$= U(\chi_E \sum a_i \chi_{E_i})$$

$$= U(\sum a_i \chi_{E \cap E_i})$$

$$= \sum a_i p_{E \cap E_i}(x)$$

$$= \sum a_i p_E(p_{E_i}(x))$$

$$= p_E(\sum a_i p_{E_i}(x))$$

$$= p_E(U(\phi))$$

$$= p_E(y),$$

which shows that $U \circ p'_E \circ U^{-1}$ and p_E agree on a dense subspace of H, whence are equal everywhere. This completes the proof.

EXERCISE 9.6. Let $E \to p_E$ be an H-projection-valued measure. Use the Hausdorff Maximality Principle to prove that there exists a sequence $\{M_i\}$ of pairwise orthogonal closed p-invariant subspaces of H, such that $E \to p_E|_{M_i}$ has a cyclic vector for each i, and such that $H = \bigoplus M_i$.

We next take up the notion of integrals with respect to a projection-valued measure.

THEOREM 9.4. *Let p be an H-projection-valued measure on a Borel space $(S, \mathcal{B})$. Let ϕ be a $\mathcal{B}$-measurable simple function, and suppose that*

$$\phi = \sum a_i \chi_{E_i} = \sum b_j \chi_{F_j},$$

where each E_i and F_j are elements of $\mathcal{B}$ and each a_i and b_j are complex numbers. Then

$$\sum a_i p_{E_i} = \sum b_j p_{F_j}.$$

EXERCISE 9.7. Prove Theorem 9.4.

DEFINITION. If p is an H-projection-valued measure on a Borel space $(S, \mathcal{B})$, and ϕ is a $\mathcal{B}$-measurable simple function on S, we define an operator, which we denote by $\int \phi \, dp$, on H by

$$\int \phi \, dp = \sum a_i p_{E_i},$$

where $\phi = \sum a_i \chi_{E_i}$. This operator is well-defined in view of the preceding theorem.

THEOREM 9.5. *Let p be an H-projection-valued measure on a Borel space $(S, \mathcal{B})$, and let X denote the space of all $\mathcal{B}$-measurable simple functions on S. Then the map L that sends ϕ to $\int \phi \, dp$ has the following properties:*

(1) $L(\phi) = \int \phi \, dp$ *is a bounded operator on H, and*

$$\|L(\phi)\| = \left\| \int \phi \, dp \right\| = \|\phi\|_\infty.$$

(2) *L is linear, i.e.,*

$$\int (\phi + \psi) \, dp = \int \phi \, dp + \int \psi \, dp$$

and

$$\int \lambda \phi \, dp = \lambda \int \phi \, dp$$

for all complex numbers λ and all $\phi, \psi \in X$.

(3) *L is multiplicative, i.e.,*

$$\int (\phi\psi)\, dp = \int \phi\, dp \circ \int \psi\, dp$$

for all $\phi, \psi \in X$.

(4) *L is essentially 1-1, i.e., $\int \phi\, dp = \int \psi\, dp$ if and only if $\phi = \psi$ a.e.p.*

(5) *For each $\phi \in X$, we have*

$$\left(\int \phi\, dp \right)^* = \int \overline{\phi}\, dp,$$

whence $\int \phi\, dp$ is selfadjoint if and only if ϕ is real-valued a.e.p.

(6) *$\int \phi\, dp$ is a positive operator if and only if ϕ is nonnegative a.e.p.*

(7) *$\int \phi\, dp$ is unitary if and only if $|\phi| = 1$ a.e.p.*

(8) *$\int \phi\, dp$ is a projection if and only if $\phi^2 = \phi$ a.e.p, i.e., if and only if ϕ agrees with a characteristic function a.e.p.*

PROOF. Let x and y be unit vectors in H, and let $\mu_{x,y}$ be the complex measure on S defined in part c of Exercise 9.2. Then

$$|([\int \phi\, dp](x), y)| = |(\sum a_i p_{E_i}(x), y)|$$

$$= |\sum a_i \mu_{x,y}(E_i)|$$

$$= |\int \phi\, d\mu_{x,y}|$$

$$\leq \|\phi\|_\infty \|\mu_{x,y}\|$$

$$\leq \|\phi\|_\infty,$$

showing that $\int \phi\, dp$ is a bounded operator and that $\|\int \phi\, dp\| \leq \|\phi\|_\infty$. See part c of Exercise 9.2 and part c of Exercise 5.12. On the other hand, we may assume that the sets $\{E_i\}$ are pairwise disjoint, that $p_{E_1} \neq 0$, and that $|a_1| = \|\phi\|_\infty$. Choosing x to be any unit vector in the range of p_{E_1}, we see that

$$[\int \phi\, dp](x) = \sum a_i p_{E_i}(p_{E_1}(x))$$

$$= \sum a_i p_{E_i \cap E_1}(x)$$

$$= a_1 p_{E_1}(x)$$

$$= a_1 x,$$

showing that $\|[\int \phi \, dp](x)\| = \|\phi\|_\infty$, and this finishes the proof of part 1.

Part 2 is left to the exercises.

To see part 3, write $\phi = \sum_{i=1}^n a_i \chi_{E_i}$, and $\psi = \sum_{j=1}^m b_j \chi_{F_j}$. Then

$$
\begin{aligned}
\int \phi \psi \, dp &= \int (\sum_{i=1}^n \sum_{j=1}^m a_i b_j \chi_{E_i} \chi_{F_j}) \, dp \\
&= \int (\sum_{i=1}^n \sum_{j=1}^m a_i b_j \chi_{E_i \cap F_j}) \, dp \\
&= \sum_{i=1}^n \sum_{j=1}^m a_i b_j \, p_{E_i \cap F_j} \\
&= \sum_{i=1}^n \sum_{j=1}^m a_i b_j \, p_{E_i} p_{F_j} \\
&= \sum_{i=1}^n a_i p_{E_i} \circ \sum_{j=1}^m b_j p_{F_j} \\
&= \int \phi \, dp \circ \int \psi \, dp,
\end{aligned}
$$

proving part 3.

We have next that $\int \phi \, dp = \int \psi \, dp$ if and only if

$$
([\int \phi \, dp](x), x) = ([\int \psi \, dp](x), x)
$$

for every $x \in H$. Therefore $\int \phi \, dp = \int \psi \, dp$ if and only if $\int \phi \, d\mu_x = \int \psi \, d\mu_x$ for every $x \in H$. If $\phi = \psi$ a.e.p, then $\phi = \psi$ a.e.μ_x for every $x \in H$, whence $\int \phi \, d\mu_x = \int \psi \, d\mu_x$ for all x, and $\int \phi \, dp = \int \psi \, dp$. Conversely, if ϕ and ψ are not equal a.e.p, then, without loss of generality, we may assume that there exists a set $E \subseteq S$ and a $\delta > 0$ such that $\phi(s) - \psi(s) > \delta$ for all $s \in E$ and $p_E \neq 0$. Letting x be a unit vector in the range of the projection p_E, we have that

$$([\int \phi \, dp](x), x) - ([\int \psi \, dp](x), x) = ([\int (\phi - \psi) \, dp](x), x)$$

$$= ([\int (\phi - \psi) \, dp](p_E(x)), x)$$

$$= ([\int (\phi - \psi) \, dp][\int \chi_E \, dp](x)), x)$$

$$= ([\int (\phi - \psi)\chi_E \, dp](x), x)$$

$$= \int (\phi - \psi)\chi_E \, d\mu_x$$

$$\geq \int \delta\chi_E \, d\mu_x$$

$$= \delta \int \chi_E \, d\mu_x$$

$$= \delta(p_E(x), x)$$

$$= \delta(x, x)$$

$$> 0,$$

proving that $\int \phi \, dp \neq \int \psi \, dp$, which gives part 4.

To see part 5, let x and y be arbitrary vectors in H. Then

$$([\int \phi \, dp]^*(x), y) = (x, [\int \phi \, dp](y))$$

$$= (x, (\sum a_i p_{E_i}(y)))$$

$$= \sum \overline{a_i}(x, p_{E_i}(y))$$

$$= \sum \overline{a_i}(p_{E_i}(x), y)$$

$$= ((\sum \overline{a_i} p_{E_i})(x), y)$$

$$= ([\int \overline{\phi} \, dp](x), y).$$

Parts 6, 7, and 8 now follow from parts 4 and 5, and we leave the details to the exercises.

EXERCISE 9.8. Prove parts 2,6,7, and 8 of Theorem 9.5.

THEOREM 9.6. Let p be an H-projection-valued measure on a Borel space $(S, \mathcal{B})$. Then the map $\phi \to L(\phi) = \int \phi \, dp$, of the space X of all $\mathcal{B}$-measurable simple functions on S into $B(H)$, extends uniquely to a map (also called L) of $L^\infty(p)$ into $B(H)$ that satisfies:

(1) L is linear.
(2) L is multiplicative; i.e., $L(fg) = L(f)L(g)$ for all $f, g \in L^\infty(p)$.
(3) $\|L(f)\| = \|f\|_\infty$ for all $f \in L^\infty(p)$.

EXERCISE 9.9. (a) Prove Theorem 9.6.

(b) If M denotes the subspace of $L^\infty(p)$ consisting of the functions f for which $f = 0$ a.e.p, show that the map L of Theorem 9.6 induces an isometric isomorphism of the Banach algebra $L^\infty(p)/M$. See part e of Exercise 9.1.

DEFINITION. If $f \in L^\infty(p)$, for p an H-projection-valued measure on $(S, \mathcal{B})$, we denote the bounded operator that is the image of f under the isometry L of the preceding theorem by $\int f \, dp$ or $\int f(s) \, dp(s)$, and we call it the *integral* of f with respect to the projection-valued measure p.

EXERCISE 9.10. Verify the following properties of the integral with respect to a projection-valued measure p.

(a) Suppose $f \in L^\infty(p)$ and $x, y \in H$. Then the matrix coefficient $([\int f \, dp](x), y)$ is given by

$$([\int f \, dp](x), y) = \int f \, d\mu_{x,y},$$

where $\mu_{x,y}$ is the complex measure defined in part c of Exercise 9.2.

(b) $[\int f \, dp]^* = \int \overline{f} \, dp$, whence $\int f \, dp$ is selfadjoint if and only if f is real-valued a.e.p.

(c) $\int f \, dp$ is a unitary operator if and only if $|f| = 1$ a.e.p.

(d) $\int f \, dp$ is a positive operator if and only if f is nonnegative a.e.p.

(e) We say that an element f in $L^\infty(p)$ is *essentially* bounded away from 0 if and only if there exists a $\delta > 0$ such that

$$p_{f^{-1}(B_\delta(0))} = 0.$$

Show that $\int f \, dp$ is invertible in $B(H)$ if and only if f is essentially bounded away from zero. HINT: If f is not essentially bounded away from 0, let $\{x_n\}$ be a sequence of unit vectors for which x_n belongs to the range of the projection $p_{f^{-1}(B_{1/n}(0))}$. Show that

$$\left\| \left[\int f \, dp \right](x_n) \right\| \leq 1/n,$$

so that no inverse of $\int f \, dp$ could be bounded.

EXERCISE 9.11. Let p be a projection-valued measure on the Borel space $(S, \mathcal{B})$.

(a) Suppose there exists a point $s \in S$ for which $p_{\{s\}} \neq 0$. Show that each operator $\int f \, dp$, for $f \in L^\infty(p)$, has an eigenvector belonging to the eigenvalue $\lambda = f(s)$. Indeed, any nonzero vector in the range of $p_{\{s\}}$ will suffice.

(b) Let f be an element of $L^\infty(p)$, let λ_0 be a complex number, let $\epsilon > 0$ be given, and write $B_\epsilon(\lambda_0)$ for the ball of radius ϵ around λ_0. Define $E = f^{-1}(B_\epsilon(\lambda_0))$, and let x be a vector in the range of p_E. Show that

$$(|\lambda_0| - \epsilon)\|x\| \leq \left\| \left[\int f \, dp \right](x) \right\| \leq (|\lambda_0| + \epsilon)\|x\|.$$

More particularly, suppose f is real-valued, that $0 < a < b \leq \infty$, and let $E = f^{-1}(a, b)$. If x is in the range of p_E, show that

$$a\|x\| \leq \left\| \left[\int f \, dp \right](x) \right\| \leq b\|x\|.$$

(c) Suppose $f \in L^\infty(p)$ is such that the operator $T = \int f \, dp$ has an eigenvector with eigenvalue λ. Define $E = f^{-1}(\{\lambda\})$. Prove that $p_E \neq 0$, and show further that $x \in H$ is an eigenvector for T belonging to the eigenvalue λ if and only if x belongs to the range of p_E.

EXERCISE 9.12. Let $E \to p_E$ be the canonical projection-valued measure on $L^2(\mu)$. Verify that $\int f \, dp$ is the multiplication operator m_f for every $f \in L^\infty(p)$. HINT: Do this first for characteristic functions χ_E.

EXERCISE 9.13. (Change of Variables) Let $(S, \mathcal{B})$ and $(S', \mathcal{B}')$ be two Borel spaces, and let h be a Borel map from S into S'; i.e., h maps S into S' and $h^{-1}(E') \in \mathcal{B}$ whenever $E' \in \mathcal{B}'$. Suppose p is a projection-valued measure on $(S, \mathcal{B})$, and as in part c of Exercise 9.3 define a projection-valued measure $q = h_*(p)$ on $(S', \mathcal{B}')$ by

$$q_E = p_{h^{-1}(E)}.$$

If f is any bounded $\mathcal{B}'$-measurable function on S', show that

$$\int f \, dq = \int (f \circ h) \, dp.$$

HINT: Check this equality for characteristic functions, then simple functions, and finally bounded functions.

THEOREM 9.7. (A "Riesz" Representation Theorem) *Let Δ be a second countable compact Hausdorff space, let H be a separable Hilbert space, and let T be a continuous linear transformation from the complex normed linear space $C(\Delta)$ of all continuous complex-valued functions on Δ into $B(H)$. Assume that T satisfies*

(1) $T(fg) = T(f) \circ T(g)$ *for all* $f, g \in C(\Delta)$.
(2) $T(\bar{f}) = [T(f)]^*$ *for all* $f \in C(\Delta)$.
(3) $T(1) = I$, *where 1 denotes the identically 1 function and I denotes the identity operator on H.*

Then there exists a unique projection-valued measure $E \to p_E$ from the σ-algebra $\mathcal{B}$ of Borel subsets of Δ such that

$$T(f) = \int f \, dp$$

for every $f \in C(\Delta)$.

PROOF. Note first that assumptions 1 and 2 imply that $T(f)$ is a positive operator if $f \geq 0$. Next, for each pair (x, y) of vectors in H, define $\phi_{x,y}$ on $C(\Delta)$ by

$$\phi_{x,y}(f) = (L(f)(x), y).$$

Then $\phi_{x,y}$ is a bounded linear functional on $C(\Delta)$, and we write $\nu_{x,y}$ for the unique finite complex Borel measure on Δ for which

$$\phi_{x,y}(f) = \int f \, d\nu_{x,y}$$

for all $f \in C(\Delta)$. See Theorem 1.5 and Exercise 1.12. We see immediately that

(1) The linear functional $\phi_{x,x}$ is a positive linear functional, whence the measure $\nu_{x,x}$ is a positive measure.
(2) For each fixed $y \in H$, the map $x \to \nu_{x,y}$ is a linear transformation of H into the vector space $M(\Delta)$ of all finite complex Borel measures on Δ.
(3) $\nu_{x,y} = \overline{\nu_{y,x}}$ for all $x, y \in H$.
(4) $\|\nu_{x,y}\| = \|\phi_{x,y}\| \leq \|x\|\|y\|$.

For each bounded, real-valued, Borel function h on Δ, consider the map $L_h : H \times H \to \mathbb{C}$ given by

$$L_h(x, y) = \int h \, d\nu_{x,y}.$$

It follows from the results above that for each fixed $y \in H$ the map $x \to L_h(x, y)$ is linear. Also,

$$
\begin{aligned}
L_h(y, x) &= \int h \, d\nu_{y,x} \\
&= \int h \, d\bar{\nu}_{y,x} \\
&= \overline{\int \bar{h} \, d\bar{\nu}_{y,x}} \\
&= \overline{\int h \, d\nu_{x,y}} \\
&= \overline{L_h(x, y)}
\end{aligned}
$$

for all $x, y \in H$. Furthermore, using Exercise 5.12 we have that

$$
\begin{aligned}
|L_h(x, y)| &= |\int h \, d\nu_{x,y}| \\
&\leq \|h\|_\infty \|\nu_{x,y}\| \\
&\leq \|h\|_\infty \|x\| \|y\|.
\end{aligned}
$$

Now, using Theorem 8.5, let $T(h)$ be the unique bounded operator on H for which

$$L_h(x, y) = (T(h)(x), y)$$

for all $x, y \in H$. Note that since the measures $\nu_{x,x}$ are positive measures, it follows that the matrix coefficients

$$(T(h)(x), x) = L_h(x, x) = \int h \, d\nu_{x,x}$$

are all real, implying that the operator $T(h)$ is selfadjoint.

If E is a Borel subset of Δ, set $p_E = T(\chi_E)$. We will eventually see that the assignment $E \to p_E$ is a projection-valued measure on $(\Delta, \mathcal{B})$.

Fix $g \in C(\Delta)$ and $x, y \in H$. Note that the two bounded linear functionals

$$f \to \int f g \, d\nu_{x,y} = \phi_{x,y}(fg) = (T(fg)(x), y)$$

and

$$f \to \int f \, d\nu_{T(g)(x),y} = \phi_{T(g)(x),y}(f) = (T(f)(T(g)(x)), y)$$

agree on $C(\Delta)$. Since they are both represented by integrals (Theorem 1.5), it follows that

$$\int hg \, d\nu_{x,y} = \int h \, d\nu_{T(g)(x),y}$$

for every bounded Borel function h. Now, for each fixed bounded, real-valued, Borel function h and each pair $x, y \in H$, the two bounded linear functionals

$$g \to \int gh \, d\nu_{x,y} = \int hg \, d\nu_{x,y}$$

and

$$g \to \int h \, d\nu_{T(g)(x),y} = (T(h)(T(g)(x)), y)$$
$$= (T(g)(x), T(h)(y))$$
$$= \int g \, d\nu_{x,T(h)(y)}$$

agree on $C(\Delta)$. Again, since both functionals can be represented as integrals, it follows that

$$\int hk \, d\nu_{x,y} = \int k \, d\nu_{x,T(h)(y)}$$

for all bounded, real-valued, Borel functions h and k. Therefore,

$$(T(hk)(x), y) = L_{hk}(x, y)$$
$$= \int hk \, d\nu_{x,y}$$
$$= \int k \, d\nu_{x,T(h)(y)}$$
$$= L_k(x, T(h)(y))$$
$$= (T(k)(x), T(h)(y))$$
$$= (T(h)(T(k)(x)), y),$$

showing that $T(hk) = T(h)T(k)$ for all bounded, real-valued, Borel functions h and k.

We see directly from the preceding calculation that each $p_E = T(\chi_E)$ is a projection. Clearly $p_\Delta = T(1) = I$ and $p_\emptyset = T(0) = 0$, so that to see that $E \to p_E$ is a projection-valued measure we must only check the countable additivity condition. Thus, let $\{E_n\}$ be a sequence of pairwise disjoint Borel subsets of Δ, and write $E = \cup E_n$. For any vectors $x, y \in H$, we have

$$(p_E(x), y) = (T(\chi_E)(x), y)$$
$$= L_{\chi_E}(x, y)$$
$$= \int \chi_E \, d\nu_{x,y}$$
$$= \nu_{x,y}(E)$$
$$= \sum \nu_{x,y}(E_n)$$
$$= \sum (p_{E_n}(x), y)$$
$$= ([\sum p_{E_n}](x), y),$$

as desired.

Finally, let us show that $T(f) = \int f \, dp$ for every $f \in C(\Delta)$. Note that, for vectors $x, y \in H$, we have that the measure $\nu_{x,y}$ agrees with the measure $\mu_{x,y}$, where $\mu_{x,y}$ is the measure defined in part c of Exercise 9.2 by

$$\mu_{x,y}(E) = (p_E(x), y).$$

We then have

$$(T(f)(x), y) = \phi_{x,y}(f)$$
$$= \int f \, d\nu_{x,y}$$
$$= \int f \, d\mu_{x,y}$$
$$= ([\int f \, dp](x), y),$$

by part a of Exercise 9.10. This shows the desired equality of $T(f)$ and $\int f \, dp$.

The uniqueness of the projection-valued measure p, satisfying $T(f) = \int f \, dp$ for all $f \in C(\Delta)$, follows from part d of Exercise 9.2.

We close this chapter by attempting to extend the definition of integral with respect to a projection-valued measure to unbounded measurable functions. For simplicity, we will restrict our attention to real-valued functions.

DEFINITION. Let p be an H-projection-valued measure on the Borel space $(S, \mathcal{B})$, and let f be a real-valued, $\mathcal{B}$-measurable function on S. For each integer n, define $E_n = f^{-1}(-n, n)$, and write T_n for the bounded selfadjoint operator on H given by $T_n = \int f \chi_{E_n} \, dp$. We define $D(f)$ to be the set of all $x \in H$ for which $\lim_n T_n(x)$ exists, and we define $T_f : D(f) \to H$ by $T_f(x) = \lim T_n(x)$.

EXERCISE 9.14. Using the notation of the preceding definition, show that
(a) If x is in the range of p_{E_n}, then $x \in D(f)$, and $T_f(x) = T_n(x)$.
(b) $x \in D(f)$ if and only if the sequence $\{T_n(x)\}$ is bounded. HINT: $x = p_{E_n}(x) + p_{\tilde{E}_n}(x)$. Show further that the sequence $\{\|T_n(x)\|\}$ is increasing.
(c) $D(f)$ is a subspace of H and T_f is a linear transformation of $D(f)$ into H.

THEOREM 9.8. *Let the notation be as in the preceding definition.*
(1) $D(f)$ *is a dense subspace of H.*
(2) T_f *is symmetric on $D(f)$; i.e.,*

$$(T_f(x), y) = (x, T_f(y))$$

for all $x, y \in D(f)$.
(3) *The graph of T_f is a closed subspace in $H \times H$.*
(4) *The following are equivalent: i) $D(f) = H$; ii) T_f is continuous from $D(f)$ into H; iii) $f \in L^{\infty}(p)$.*
(5) *The linear transformations $I \pm iT_f$ are both 1-1 and onto from $D(f)$ to H.*
(6) *The linear transformation $U_f = (I - iT_f)(I + iT_f)^{-1}$ is 1-1 and onto from H to H and is in fact a unitary operator for which -1 is not an eigenvalue. (This operator U_f is called the Cayley transform of T_f.)*
(7) *The range of $I + U_f$ equals $D(f)$, and*

$$T_f = -i(I - U_f)(I + U_f)^{-1}.$$

PROOF. That $D(f)$ is dense in H follows from part a of Exercise 9.14 and part c of Exercise 9.1.
Each operator T_n is selfadjoint. So, if $x, y \in D(f)$, then

$$(T_f(x), y) = \lim(T_n(x), y) = \lim(x, T_n(y)) = (x, T_f(y)),$$

showing that T_f is symmetric on its domain $D(f)$.

The graph of T_f, like the graph of any linear transformation of H into itself, is clearly a subspace of $H \times H$. To see that the graph of T_f is closed, let (x, y) be in the closure of the graph, i.e., $x = \lim x_j$ and $y = \lim T_f(x_j)$, where each $x_j \in D(f)$. We must show that $x \in D(f)$ and then that $y = T_f(x)$. Now the sequence $\{T_f(x_j)\}$ is bounded in norm, and for each n we have from the preceding exercise that $\|T_n(x_j)\| \leq \|T_f(x_j)\|$. Hence, there exists a constant M such that $\|T_n(x_j)\| \leq M$ for all n and j. Writing $T_n(x) = T_n(x - x_j) + T_n(x_j)$, we have that

$$\|T_n(x)\| \leq \lim_j \|T_n(x - x_j)\| + M = M$$

for all n, whence $x \in D(f)$ by Exercise 9.14. Now, for any $z \in D(f)$ we have

$$\begin{aligned}(y, z) &= \lim(T_f(x_j), z) \\ &= \lim(x_j, T_f(z)) \\ &= (x, T_f(z)) \\ &= (T_f(x), z),\end{aligned}$$

proving that $y = T_f(x)$ since $D(f)$ is dense in H.

We prove part 4 by showing that i) implies ii), ii) implies iii), and iii) implies i). First, if $D(f) = H$, then by the Closed Graph Theorem we have that T_f is continuous. Next, if f is not an element of $L^\infty(p)$, then there exists an increasing sequence $\{n_k\}$ of positive integers for which either

$$p_{f^{-1}(n_k, n_{k+1})} \neq 0$$

for all k, or

$$p_{f^{-1}(-n_{k+1}, -n_k)} \neq 0$$

for all k. Without loss of generality, suppose that

$$p_{f^{-1}(n_k, n_{k+1})} \neq 0$$

for all k. Write $F_k = f^{-1}(n_k, n_{k+1})$, and note that $F_k \subseteq E_{n_{k+1}}$. Now, for each k, let x_k be a unit vector in the range of p_{F_k}. Then each $x_k \in D(f)$, and

$$\begin{aligned}(T_f(x_k), x_k) &= (T_{n_{k+1}}(x_k), x_k) \\ &= ((T_{n_{k+1}} \circ p_{F_k})(x_k), x_k) \\ &= \int f \chi_{F_k} \, d\mu_{x_k} \\ &\geq n_k \|x_k\|^2 \\ &= n_k,\end{aligned}$$

proving that $\|T_f(x_k)\| \geq n_k$, whence T_f is not continuous. Finally, if $f \in L^\infty(p)$, then clearly $T_f = T_n$ for any $n \geq \|f\|_\infty$, and $D(f) = H$. This proves part 4.

We show part 5 for $I + iT_f$. An analogous argument works for $I - iT_f$. Observe that, for $x \in D(f)$, we have

$$\|(I + iT_f)(x)\|^2 = ((I + iT_f)(x), (I + iT_f)(x)) = \|x\|^2 + \|T_f(x)\|^2.$$

Therefore, $I + iT_f$ is norm-increasing, whence is 1-1. Now, if $\{(I + iT_f)(x_j)\}$ is a sequence of elements in the range of $I + iT_f$ that converges to a point $y \in H$, then the sequence $\{(I + iT_f)(x_j)\}$ is a Cauchy sequence and therefore, since $I + iT_f$ is norm-increasing, the sequence $\{x_j\}$ is a Cauchy sequence as well. Let $x = \lim_j x_j$. It follows that $y = x + iz$, where $z = \lim_j T_f(x_j)$. Since the graph of T_f is closed, we must have that $x \in D(f)$ and $z = T_f(x)$. Hence, $y = (I + iT_f)(x)$ belongs to the range of $I + iT_f$, showing that this range is closed. We complete the proof then of part 5 by showing that the range of $I + iT_f$ is dense in H. Thus, if $y \in H$ is orthogonal to every element of the range of $I + iT_f$, then for each n we have

$$
\begin{aligned}
0 &= ((I + iT_f)(p_{E_n}(y)), y) \\
&= ((I + iT_f)(p_{E_n}^2(y)), y) \\
&= ((I + iT_n)(p_{E_n}(y)), y) \\
&= (p_{E_n}(I + iT_n)p_{E_n}(y), y) \\
&= ((I + iT_n)p_{E_n}(y), p_{E_n}(y)) \\
&= \|p_{E_n}(y)\|^2 + i(T_n(p_{E_n}(y)), p_{E_n}(y)) \\
&= \|p_{E_n}(y)\|^2 + i(T_n(y), y).
\end{aligned}
$$

But then $\|p_{E_n}(y)\|^2 = -i(T_n(p_{E_n}(y)), p_{E_n}(y))$, which, since T_n is self-adjoint, can happen only if $p_{E_n}(y) = 0$. But then $y = \lim_n p_{E_n}(y)$ must be 0. Therefore, the range of $I + iT_f$ is dense, whence is all of H.

Next, since $I + iT_f$ and $I - iT_f$ are both 1-1 from $D(f)$ onto H, it follows that $U_f = (I - iT_f)(I + iT_f)^{-1}$ is 1-1 from H onto itself. Further, writing $y \in D(f)$ as $(I + iT_f)^{-1}(x)$, we have

$$
\begin{aligned}
\|U_f(x)\|^2 &= \|(I - iT_f)((I + iT_f)^{-1}(x))\|^2 \\
&= \|(I - iT_f)(y)\|^2 \\
&= \|y\|^2 + \|T_f(y)\|^2 \\
&= \|(I + iT_f)(y)\|^2 \\
&= \|x\|^2,
\end{aligned}
$$

proving that U_f is unitary. Writing the identity operator I as $(I + iT_f)(I + iT_f)^{-1}$, we have that $I + U_f = 2(I + iT_f)^{-1}$, which is 1-1. Consequently, -1 is not an eigenvalue for U_f.

We leave the verification of part 7 to the exercises. This completes the proof.

DEFINITION. We call the operator $T_f : D(f) \to H$ of the preceding theorem the *integral* of f with respect to p, and we denote it by $\int f\, dp$ or $\int f(s)\, dp(s)$. It is not in general an element of $B(H)$. Indeed, as we have seen in the preceding theorem, $\int f\, dp$ is in $B(H)$ if and only if f is in $L^\infty(p)$.

EXERCISE 9.15. (a) Prove part 7 of Theorem 9.8.

(b) Suppose (S, μ) is a σ-finite measure space, that p is the canonical projection-valued measure on $L^2(\mu)$, and that f is a real-valued measurable function on S. Verify that $D(f)$ is the set of all L^2 functions g for which $fg \in L^2(\mu)$, and that $[\int f\, dp](g) = fg$ for all $g \in D(f)$.

(c) Suppose $(S, \mathcal{B})$ and p are as in the preceding theorem. Let f be a measurable real-valued function on S and let g be a bounded measurable real-valued function on S. Being careful about domains, prove that

$$[\int gf\, dp] = [\int g\, dp] \circ [\int f\, dp].$$

Show that $\int gf\, dp$ need not be the same as $[\int f\, dp] \circ [\int g\, dp]$. HINT: Use part b.

(d) Let $S, \mathcal{B}$, and p be as in part c. Suppose g is an everywhere nonzero, bounded, real-valued, measurable function on S, and write T for the bounded operator $\int g\, dp$. Prove that the operator $\int(1/g)\, dp$ is a right inverse for the operator T.

(e) Let $(S, \mathcal{B})$ and $(S', \mathcal{B}')$ be two Borel spaces, and let h be a Borel map from S into S'. Suppose p is a projection-valued measure on $(S, \mathcal{B})$, and as in part c of Exercise 9.3 define a projection-valued measure $q = h_*(p)$ on $(S', \mathcal{B}')$ by

$$q_E = p_{h^{-1}(E)}.$$

If f is any (possibly unbounded) real-valued $\mathcal{B}'$-measurable function on S', show that

$$\int f\, dq = \int (f \circ h)\, dp.$$

EXERCISE 9.16. Let p be the projection-valued measure on the Borel space $(\mathbb{R}, \mathcal{B})$ of part b of Exercise 9.3.

(a) Show that

$$\int f \, dp = U^{-1} \circ m_f \circ U$$

for every $f \in L^\infty(p)$.

(b) If $f(x) = x$, and $T_f = \int f \, dp$, show that $D(f)$ consists of all the L^2 functions g for which $x[U(g)](x) \in L^2(\mathbb{R})$, and then show that every such g is absolutely continuous and has an L^2 derivative.

(c) Conclude that the operator $\int f \, dp$ of part b has for its domain the set of all L^2 absolutely continuous functions having L^2 derivatives, and that $[\int f \, dp](g) = (1/2\pi i)g'$.

CHAPTER X

THE SPECTRAL THEOREM OF GELFAND

DEFINITION A *Banach algebra* is a complex Banach space A on which there is defined an associative multiplication $\times$ for which:

(1) $x \times (y + z) = x \times y + x \times z$ and $(y + z) \times x = y \times x + z \times x$ for all $x, y, z \in A$.
(2) $x \times (\lambda y) = \lambda x \times y = (\lambda x) \times y$ for all $x, y \in A$ and $\lambda \in \mathbb{C}$.
(3) $\|x \times y\| \leq \|x\| \|y\|$ for all $x, y \in A$.

We call the Banach algebra *commutative* if the multiplication in A is commutative.

An *involution* on a Banach algebra A is a map $x \to x^*$ of A into itself that satisfies the following conditions for all $x, y \in A$ and $\lambda \in \mathbb{C}$.

(1) $(x + y)^* = x^* + y^*$.
(2) $(\lambda x)^* = \overline{\lambda} x^*$.
(3) $(x^*)^* = x$.
(4) $(x \times y)^* = y^* \times x^*$.
(5) $\|x^*\| = \|x\|$.

We call x^* the *adjoint* of x. A subset $S \subseteq A$ is called *selfadjoint* if $x \in S$ implies that $x^* \in S$.

A Banach algebra A on which there is defined an involution is called a *Banach *-algebra*.

An element of a Banach *-algebra is called *selfadjoint* if $x^* = x$. If a Banach *-algebra A has an identity I, then an element $x \in A$, for which $x \times x^* = x^* \times x = I$, is called a *unitary element* of A. A selfadjoint element x, for which $x^2 = x$, is called a *projection* in A. An element x

187

that commutes with its adjoint x^* is called a *normal element* of A.

A Banach algebra A is a C^*-*algebra* if it is a Banach *-algebra, and if the equation

$$\|x \times x^*\| = \|x\|^2$$

holds for all $x \in A$. A *sub* C^*-*algebra* of a C^*-algebra A is a subalgebra B of A that is a closed subset of the Banach space A and is also closed under the adjoint operation.

REMARK. We ordinarily write xy instead of $x \times y$ for the multiplication in a Banach algebra. It should be clear that the axioms for a Banach algebra are inspired by the properties of the space $B(H)$ of bounded linear operators on a Hilbert space H.

EXERCISE 10.1. (a) Let A be the set of all $n \times n$ complex matrices, and for $M = [a_{ij}] \in A$ define

$$\|M\| = \sqrt{\sum_{i=1}^{n} \sum_{j=1}^{n} |a_{ij}|^2}.$$

Prove that A is a Banach algebra with identity I. Verify that A is a Banach *-algebra if M^* is defined to be the complex conjugate of the transpose of M. Give an example to show that A is not a C^*-algebra.

(b) Suppose H is a Hilbert space. Verify that $B(H)$ is a C^*-algebra. Using as H the Hilbert space $\mathbb{C}^2$, give an example of an element $x \in B(H)$ for which $\|x^2\| \neq \|x\|^2$. Observe that this example is not the same as that in part a. (The norms are different.)

(c) Verify that $L^1(\mathbb{R})$ is a Banach algebra, where multiplication is defined to be convolution. Show further that, if $f^*(x)$ is defined to be $\overline{f(-x)}$, then $L^1(\mathbb{R})$ is a Banach *-algebra. Give an example to show that $L^1(\mathbb{R})$ is not a C^*-algebra.

(d) Verify that $C_0(\Delta)$ is a Banach algebra, where Δ is a locally compact Hausdorff space, the algebraic operations are pointwise, and the norm on $C_0(\Delta)$ is the supremum norm. Show further that $C_0(\Delta)$ is a C^*-algebra, if we define f^* to be $\overline{f}$. Show that $C_0(\Delta)$ has an identity if and only if Δ is compact.

(e) Let A be an arbitrary Banach algebra. Prove that the map $(x, y) \to xy$ is continuous from $A \times A$ into A.

(f) Let A be a Banach algebra. Suppose $x \in A$ satisfies $\|x\| < 1$. Prove that $0 = \lim_n x^n$.

(g) Let M be a closed subspace of a Banach algebra A, and assume that M is a two-sided ideal in (the ring) A; i.e., $xy \in M$ and $yx \in M$ if $x \in A$ and $y \in M$. Prove that the Banach space A/M is a Banach algebra and that the natural map $\pi : A \to A/M$ is a continuous homomorphism of the Banach algebra A onto the Banach algebra A/M.

(h) Let A be a Banach algebra with identity I and let x be an element of A. Show that the smallest subalgebra B of A that contains x coincides with the set of all polynomials in x, i.e., the set of all elements y of the form $y = \sum_{j=0}^{n} a_j x^j$, where each a_j is a complex number and $x^0 = I$. We denote this subalgebra by $[x]$ and call it the subalgebra of A *generated by x.*

(i) Let A be a Banach *-algebra. Show that each element $x \in A$ can be written uniquely as $x = x_1 + ix_2$, where x_1 and x_2 are selfadjoint. Show further that if A contains an identity I, then $I^* = I$. If A is a C^*-algebra with identity, and if U is a unitary element in A, show that $\|U\| = 1$.

(j) Let x be a selfadjoint element of a C^*-algebra A. Prove that $\|x^n\| = \|x\|^n$ for all nonnegative integers n. HINT: Do this first for $n = 2^k$.

EXERCISE 10.2. (Adjoining an Identity) Let A be a Banach algebra, and let B be the complex vector space $A \times \mathbb{C}$. Define a multiplication on B by

$$(x, \lambda) \times (x', \lambda') = (xx' + \lambda x' + \lambda' x \, , \, \lambda\lambda'),$$

and set $\|(x, \lambda)\| = \|x\| + |\lambda|$.

(a) Prove that B is a Banach algebra with identity.

(b) Show that the map $x \to (x, 0)$ is an isometric isomorphism of the Banach algebra A onto an ideal M of B. Show that M is of codimension 1; i.e., the dimension of B/M is 1. (This map $x \to (x, 0)$ is called the *canonical isomorphism* of A into B.

(c) Conclude that every Banach algebra is isometrically isomorphic to an ideal of codimension 1 in a Banach algebra with identity.

(d) Suppose A is a Banach algebra with identity, and let B be the Banach algebra $A \times \mathbb{C}$ constructed above. What is the relationship, if any, between the identity in A and the identity in B?

(e) If A is a Banach *-algebra, can A be imbedded isometrically and isomorphically as an ideal of codimension 1 in a Banach *-algebra?

THEOREM 10.1. *Let x be an element of a Banach algebra A with identity I, and suppose that $\|x\| = \alpha < 1$. Then the element $I - x$ is* *invertible in A and*

$$(I - x)^{-1} = \sum_{n=0}^{\infty} x^n.$$

PROOF. The sequence of partial sums of the infinite series $\sum_{n=0}^{\infty} x^n$ forms a Cauchy sequence in A, for

$$\left\| \sum_{n=0}^{j} x^n - \sum_{n=0}^{k} x^n \right\| = \left\| \sum_{n=k+1}^{j} x^n \right\|$$

$$\leq \sum_{n=k+1}^{j} \|x^n\|$$

$$\leq \sum_{n=k+1}^{j} \|x\|^n$$

$$= \sum_{n=k+1}^{j} \alpha^n.$$

We write

$$y = \sum_{n=0}^{\infty} x^n = \lim_j \sum_{n=0}^{j} x^n = \lim_j S_j.$$

Then

$$(I - x)y = \lim_j (I - x)S_j$$

$$= \lim_j (I - x) \sum_{n=0}^{j} x^n$$

$$= \lim_j (I - x^{j+1})$$

$$= I,$$

by part f of Exercise 10.1, showing that y is a right inverse for $I - x$. That y also is a left inverse follows similarly, whence $y = (I - x)^{-1}$, as desired.

EXERCISE 10.3. Let A be a Banach algebra with identity I.

(a) If $x \in A$ satisfies $\|x\| < 1$, show that $I + x$ is invertible in A.

(b) Suppose $y \in A$ is invertible, and set $\delta = 1/\|y^{-1}\|$. Prove that x is invertible in A if $\|x - y\| < \delta$. HINT: Write $x = y(I + y^{-1}(x - y))$.

(c) Conclude that the set of invertible elements in A is a nonempty, proper, open subset of A.

(d) Prove that the map $x \to x^{-1}$ is continuous on its domain. HINT: $y^{-1} - x^{-1} = y^{-1}(x - y)x^{-1}$.

(e) Let x be an element of A. Show that the infinite series

$$\sum_{n=0}^{\infty} x^n/n!$$

converges to an element of A. Define

$$e^x = \sum_{n=0}^{\infty} x^n/n!.$$

Show that

$$e^{x+y} = e^x e^y$$

if $xy = yx$. Compare with part c of Exercise 8.13.

(f) Suppose in addition that A is a Banach *-algebra and that x is a selfadjoint element of A. Prove that e^{ix} is a unitary element of A. Compare with part d of Exercise 8.13.

THEOREM 10.2. (Mazur's Theorem) *Let A be a Banach algebra with identity I, and assume further that A is a division ring, i.e., that every nonzero element of A has a multiplicative inverse. Then A consists of the complex multiples λI of the identity I, and the map $\lambda \to \lambda I$ is a topological isomorphism of $\mathbb{C}$ onto A.*

PROOF. Assume false, and let x be an element of A that is not a complex multiple of I. This means that each element $x_\lambda = x - \lambda I$ has an inverse.

Let f be an arbitrary element of the conjugate space A^* of A, and define a function F of a complex variable λ by

$$F(\lambda) = f(x_\lambda^{-1}) = f((x - \lambda I)^{-1}).$$

We claim first that F is an entire function of λ. Thus, let λ be fixed. We use the factorization formula

$$y^{-1} - z^{-1} = y^{-1}(z - y)z^{-1}.$$

We have

$$F(\lambda + h) - F(\lambda) = f(x_{\lambda+h}^{-1}) - f(x_\lambda^{-1})$$
$$= f(x_{\lambda+h}^{-1}(x_\lambda - x_{\lambda+h})x_\lambda^{-1})$$
$$= -hf(x_{\lambda+h}^{-1}x_\lambda^{-1}).$$

So,

$$\lim_{h \to 0} \frac{F(\lambda + h) - F(\lambda)}{h} = -f(x_\lambda^{-2}),$$

and F is differentiable everywhere. See part d of Exercise 10.3.

Next, observe that

$$\lim_{\lambda \to \infty} F(\lambda) = \lim_{\lambda \to \infty} f((x - \lambda I)^{-1})$$
$$= \lim_{\lambda \to \infty} (1/\lambda)f(((x/\lambda) - I)^{-1})$$
$$= 0.$$

Therefore, F is a bounded entire function, and so by Liouville's Theorem, $F(\lambda) = 0$ identically. Consequently, $f(x_0^{-1}) = f(x^{-1}) = 0$ for all $f \in A^*$. But this would imply that $x^{-1} = 0$, which is a contradiction.

We introduce next a dual object for Banach algebras that is analogous to the conjugate space of a Banach space.

DEFINITION. Let A be a Banach algebra. By the *structure space* of A we mean the set Δ of all nonzero continuous algebra homomorphisms (linear and multiplicative) $\phi : A \to \mathbb{C}$. The structure space is a (possibly empty) subset of the conjugate space A^*, and we think of Δ as being equipped with the inherited weak* topology.

THEOREM 10.3. *Let A be a Banach algebra, and let Δ denote its structure space. Then Δ is locally compact and Hausdorff. Further, if A is a separable Banach algebra, then Δ is second countable and metrizable. If A contains an identity I, then Δ is compact.*

PROOF. Δ is clearly a Hausdorff space since the weak* topology on A^* is Hausdorff.

Observe next that if $\phi \in \Delta$, then $\|\phi\| \leq 1$. Indeed, for any $x \in A$, we have

$$|\phi(x)| = |\phi(x^n)|^{1/n} \leq \|\phi\|^{1/n}\|x\| \to \|x\|,$$

implying that $\|\phi\| \leq 1$, as claimed. It follows then that Δ is contained in the closed unit ball $\overline{B_1}$ of A^*. Since the ball $\overline{B_1}$ in A^* is by Alaoglu's

Theorem compact in the weak* topology, we could show that Δ is compact by verifying that it is closed in $\overline{B_1}$. This we can do if A contains an identity I. Thus, let $\{\phi_\alpha\}$ be a net of elements of Δ that converges in the weak* topology to an element $\phi \in \overline{B_1}$. Since this convergence is pointwise convergence on A, it follows that $\phi(xy) = \phi(x)\phi(y)$, for all $x, y \in A$, whence ϕ is a homomorphism of the algebra A into $\mathbb{C}$. Also, since every nonzero homomorphism of A must map I to 1, it follows that $\phi(I) = 1$, whence ϕ is not the 0 homomorphism. Hence, $\phi \in \Delta$, as desired.

We leave the proof that Δ is always locally compact to the exercises.

Of course, if A is separable, then the weak* topology on $\overline{B_1}$ is compact and metrizable, so that Δ is second countable and metrizable in this case, as desired.

EXERCISE 10.4. Let A be a Banach algebra.

(a) Suppose that the elements of the structure space Δ of A separate the points of A. Prove that A is commutative.

(b) Suppose A is the algebra of all $n \times n$ complex matrices as defined in part a of Exercise 10.1. Prove that the structure space Δ of A is the empty set if $n > 1$.

(c) If A has no identity, show that Δ is locally compact. HINT: Show that the closure of Δ in $\overline{B_1}$ is contained in the union of Δ and $\{0\}$, whence Δ is an open subset of a compact Hausdorff space.

(d) Let B be the Banach algebra with identity constructed from A as in Exercise 10.2, and identify A with its canonical isomorphic image in B. Prove that every element ϕ in the structure space Δ_A of A has a unique extension to an element ϕ' in the structure space Δ_B of B. Show that there exists a unique element $\phi_0 \in \Delta_B$ whose restriction to A is identically 0. Show further that the above map $\phi \to \phi'$ is a homeomorphism of Δ_A onto $\Delta_B - \{\phi_0\}$.

DEFINITION. Let A be a Banach algebra and let Δ be its structure space. For each $x \in A$, define a function $\hat{x}$ on Δ by

$$\hat{x}(\phi) = \phi(x).$$

The map $x \to \hat{x}$ is called the *Gelfand transform* of A, and the function $\hat{x}$ is called the *Gelfand transform of x.*

EXERCISE 10.5. Let A be the Banach algebra $L^1(\mathbb{R})$ of part c of Exercise 10.1, and let Δ be its structure space.

(a) If λ is any real number, define $\phi_\lambda : A \to \mathbb{C}$ by

$$\phi_\lambda(f) = \int f(x)e^{-2\pi i \lambda x}\, dx.$$

Show that ϕ_λ is an element of Δ.

(b) Let ϕ be an element of Δ, and let h be the L^∞ function satisfying

$$\phi(f) = \int f(x)\overline{h(x)}\, dx.$$

Prove that $h(x+y) = h(x)h(y)$ for almost all pairs $(x, y) \in \mathbb{R}^2$. HINT: Show that

$$\int\int f(x)g(y)\overline{h}(x+y)\, dy dx = \int\int f(x)g(y)\overline{h(x)}\,\overline{h(y)}\, dy dx$$

for all $f, g \in L^1(\mathbb{R})$.

(c) Let ϕ and h be as in part b, and let f be an element of $L^1(\mathbb{R})$ for which $\phi(f) \neq 0$. Write f_x for the function defined by $f_x(y) = f(x+y)$. Show that the map $x \to \phi(f_x)$ is continuous, and that

$$h(x) = \overline{\phi(f_{-x})/\phi(f)}$$

for almost all x. Conclude that h may be chosen to be a continuous function in $L^\infty(\mathbb{R})$, in which case $h(x+y) = h(x)h(y)$ for all $x, y \in \mathbb{R}$.

(d) Suppose h is a bounded continuous map of $\mathbb{R}$ into $\mathbb{C}$, which is not identically 0 and which satisfies $h(x+y) = h(x)h(y)$ for all x and y. Show that there exists a real number λ such that $h(x) = e^{2\pi i \lambda x}$ for all x. HINT: If h is not identically 1, show that there exists a smallest positive number δ for which $h(\delta) = 1$. Show then that $h(\delta/2) = -1$ and $h(\delta/4) = \pm i$. Conclude that $\lambda = \pm(1/\delta)$ depending on whether $h(\delta/4) = i$ or $-i$.

(e) Conclude that the map $\lambda \to \phi_\lambda$ of part a is a homeomorphism between $\mathbb{R}$ and the structure space Δ of $L^1(\mathbb{R})$. HINT: To prove that the inverse map is continuous, suppose that $\{\lambda_n\}$ does not converge to λ. Show that there exists an $f \in L^1(\mathbb{R})$ such that $\int f(x)e^{-2\pi i \lambda_n x}\, dx$ does not approach $\int f(x)e^{-2\pi i \lambda x}\, dx$.

(f) Show that, using the identification of Δ with $\mathbb{R}$ in part e, that the Gelfand transform on $L^1(\mathbb{R})$ and the Fourier transform on $L^1(\mathbb{R})$ are identical. Conclude that the Gelfand transform is 1-1 on $L^1(\mathbb{R})$.

THEOREM 10.4. *Let A be a Banach algebra. Then the Gelfand transform of A is a norm-decreasing homomorphism of A into the Banach algebra $C(\Delta)$ of all continuous complex-valued functions on Δ.*

EXERCISE 10.6. (a) Prove Theorem 10.4.

(b) If A is a Banach algebra without an identity, show that each function $\hat{x}$ in the range of the Gelfand transform is an element of $C_0(\Delta)$. HINT: The closure of Δ in $\overline{B_1}$ is contained in the union of Δ and $\{0\}$.

DEFINITION. Let A be a Banach algebra with identity I, and let x be an element of A. By the *resolvent* of x we mean the set $\mathrm{res}_A(x)$ of all complex numbers λ for which $\lambda I - x$ has an inverse in A. By the *spectrum* $\mathrm{sp}_A(x)$ of x we mean the complement of the resolvent of x; i.e., $\mathrm{sp}_A(x)$ is the set of all $\lambda \in \mathbb{C}$ for which $\lambda I - x$ does not have an inverse in A. We write simply $\mathrm{res}(x)$ and $\mathrm{sp}(x)$ when it is unambiguous what the algebra A is.

By the *spectral radius* (relative to A) of x we mean the extended real number $\|x\|_{\mathrm{sp}}$ defined by

$$\|x\|_{\mathrm{sp}} = \sup_{\lambda \in \mathrm{sp}_A(x)} |\lambda|.$$

EXERCISE 10.7. Let A be a Banach algebra with identity I, and let x be an element of A.

(a) Show that the resolvent $\mathrm{res}_A(x)$ of x is open in $\mathbb{C}$, whence the spectrum $\mathrm{sp}_A(x)$ of x is closed.

(b) Show that the spectrum of x is nonempty, whence the spectral radius of x is nonnegative. HINT: Make an argument similar to the proof of Mazur's theorem.

(c) Show that $\|x\|_{\mathrm{sp}} \leq \|x\|$, whence the spectrum of x is compact. HINT: If $\lambda \neq 0$, then $\lambda I - x = (1/\lambda)(I - (x/\lambda))$.

(d) Show that there exists a complex number λ such that $\|x\|_{\mathrm{sp}} = |\lambda|$, i.e., the spectral radius is attained.

(e) (Spectral Mapping Theorem) If $p(z)$ is any complex polynomial, show that

$$\mathrm{sp}_A(p(x)) = p(\mathrm{sp}_A(x));$$

i.e., $\mu \in \mathrm{sp}_A(p(x))$ if and only if there exists a $\lambda \in \mathrm{sp}_A(x)$ such that $\mu = p(\lambda)$. HINT: Factor the polynomial $p(z) - \mu$ as

$$p(z) - \mu = c \prod_{i=1}^{n} (z - \lambda_i),$$

whence

$$p(x) - \mu I = c \prod_{i=1}^{n} (x - \lambda_i I).$$

Now, the left hand side fails to have an inverse if and only if some one of the factors on the right hand side fails to have an inverse.

THEOREM 10.5. *Let A be a commutative Banach algebra with identity I, and let x be an element of A. Then the spectrum $\mathrm{sp}_A(x)$ of x coincides with the range of the Gelfand transform $\hat{x}$ of x. Consequently, we have*

$$\|x\|_{\mathrm{sp}} = \|\hat{x}\|_{\infty}.$$

PROOF. If there exists a ϕ in the structure space Δ of A for which $\hat{x}(\phi) = \lambda$, then

$$\phi(\lambda I - x) = \lambda - \phi(x) = \lambda - \hat{x}(\phi) = 0,$$

from which it follows that $\lambda I - x$ cannot have an inverse. Hence, the range of $\hat{x}$ is contained in $\mathrm{sp}(x)$.

Conversely, let λ be in the spectrum of x. Let J be the set of all multiples $(\lambda I - x)y$ of $\lambda I - x$ by elements of A. Then J is an ideal in A, and it is a proper ideal since $\lambda I - x$ has no inverse (I is not in J). By Zorn's Lemma, there exists a maximal proper ideal M containing J. Now the closure of M is an ideal. If this closure of M is all of A, then there must exist a sequence $\{m_n\}$ of elements of M that converges to I. But, since the set of invertible elements in A is an open set, it must be that some m_n is invertible. But then M would not be a proper ideal. Therefore, $\overline{M}$ is proper, and since M is maximal it follows that M is itself closed.

Now A/M is a Banach algebra by part g of Exercise 10.1. Also, since M is maximal, we have that A/M is a field. By Mazur's Theorem (Theorem 10.2), we have that A/M is topologically isomorphic to the set of complex numbers. The natural map $\pi : A \to A/M$ is then a continuous nonzero homomorphism of A onto $\mathbb{C}$, i.e., π is an element of Δ. Further, $\pi(\lambda I - x) = 0$ since $\lambda I - x \in J \subseteq M$. Hence, $\hat{x}(\pi) = \lambda$, showing that λ belongs to the range of $\hat{x}$.

EXERCISE 10.8. Suppose A is a commutative Banach algebra with identity I, and let Δ be its structure space. Assume that x is an element of A for which the subalgebra $[x]$ generated by x is dense in A. (See part h of Exercise 10.1.) Prove that $\hat{x}$ is a homeomorphism of Δ onto the spectrum $\mathrm{sp}_a(x)$ of x.

THEOREM 10.6. *Let A be a commutative C^*-algebra with identity I. Then, for each $x \in A$, we have $\hat{x}^* = \overline{\hat{x}}$.*

PROOF. The theorem will follow if we show that $\hat{x}$ is real-valued if x is selfadjoint. (Why?) Thus, if x is selfadjoint, and if $U = e^{ix} = \sum_{n=0}^{\infty}(ix)^n/n!$, then we have seen in part f of Exercise 10.2 and part i of Exercise 10.1 that U is unitary and that $\|U\| = \|U^{-1}\| = 1$. Therefore, if ϕ is an element of the structure space Δ of A, then $|\phi(U)| \le 1$ and $1/|\phi(U)| = |\phi(U^{-1})| \le 1$, and this implies that $|\phi(U)| = 1$. On the other hand,

$$\phi(U) = \sum_{n=0}^{\infty}(i\phi(x))^n/n! = e^{i\phi(x)}.$$

But $|e^{it}| = 1$ if and only if t is real. Hence, $\hat{x}(\phi) = \phi(x)$ is real for every $\phi \in \Delta$.

The next result is an immediate consequence of the preceding theorem.

THEOREM 10.7. *If x is a selfadjoint element of a commutative C^*-algebra A with identity, then the spectrum $\mathrm{sp}_A(x)$ of x is contained in the set of real numbers.*

EXERCISE 10.9. (A Formula for the Spectral Radius) Let A be a Banach algebra with identity I, and let x be an element of A. Write $\mathrm{sp}(x)$ for $\mathrm{sp}_A(x)$.

(a) If n is any positive integer, show that $\mu \in \mathrm{sp}(x^n)$ if and only if there exists a $\lambda \in \mathrm{sp}(x)$ such that $\mu = \lambda^n$, whence

$$\|x\|_{\mathrm{sp}} = \|x^n\|_{\mathrm{sp}}^{1/n}.$$

Conclude that

$$\|x\|_{\mathrm{sp}} \le \liminf \|x^n\|^{1/n}.$$

(b) If f is an element of A^*, show that the function $\lambda \to f((\lambda I - x)^{-1})$ is analytic on the (open) resolvent $\mathrm{res}(x)$ of x. Show that the resolvent contains all λ for which $|\lambda| > \|x\|_{\mathrm{sp}}$.

(c) Let f be in A^*. Show that the function $F(\mu) = \mu f((I - \mu x)^{-1})$ is analytic on the disk of radius $1/\|x\|_{\mathrm{sp}}$ around 0 in $\mathbb{C}$. Show further that

$$F(\mu) = \sum_{n=0}^{\infty} f(x^n)\mu^{n+1}$$

on the disk of radius $1/\|x\|$ and hence also on the (possibly) larger disk of radius $1/\|x\|_{\mathrm{sp}}$.

(d) Using the Uniform Boundedness Principle, show that if $|\mu| < 1/\|x\|_{\mathrm{sp}}$, then the sequence $\{\mu^{n+1} x^n\}$ is bounded in norm, whence

$$\limsup \|x^n\|^{1/n} \leq 1/|\mu|$$

for all such μ. Show that this implies that

$$\limsup \|x^n\|^{1/n} \leq \|x\|_{\mathrm{sp}}.$$

(e) Derive the spectral radius formula:

$$\|x\|_{\mathrm{sp}} = \lim \|x^n\|^{1/n}.$$

(f) Suppose that A is a C^*-algebra and that x is a selfadjoint element of A. Prove that
$$\|x\| = \sup_{\lambda \in \mathrm{sp}(x)} |\lambda| = \|x\|_{\mathrm{sp}}.$$

THEOREM 10.8. (Gelfand's Theorem) *Let A be a commutative C^*-algebra with identity I. Then the Gelfand transform is an isometric isomorphism of the Banach algebra A onto $C(\Delta)$, where Δ is the structure space of A.*

PROOF. We have already seen that $x \to \hat{x}$ is a norm-decreasing homomorphism of A into $C(\Delta)$. We must show that the transform is an isometry and is onto.

Now it follows from part f of Exercise 10.9 and Theorem 10.4 that $\|x\| = \|\hat{x}\|_\infty$ whenever x is selfadjoint. For an arbitrary x, write $y = x^* x$. Then
$$\|x\| = \sqrt{\|y\|}$$
$$= \sqrt{\|\hat{y}\|_\infty}$$
$$= \sqrt{\|\widehat{x^* x}\|_\infty}$$
$$= \sqrt{\|\widehat{x^*}\hat{x}\|_\infty}$$
$$= \sqrt{\||\hat{x}|^2\|_\infty}$$
$$= \sqrt{\|\hat{x}\|_\infty^2}$$
$$= \|\hat{x}\|_\infty,$$

showing that the Gelfand transform is an isometry.

By Theorem 10.6, we see that the range $\hat{A}$ of the Gelfand transform is a subalgebra of $C(\Delta)$ that separates the points of Δ and is closed under complex conjugation. Then, by the Stone-Weierstrass Theorem, $\hat{A}$ must be dense in $C(\Delta)$. But, since A is itself complete, and the Gelfand transform is an isometry, it follows that $\hat{A}$ is closed in $C(\Delta)$, whence is all of $C(\Delta)$.

EXERCISE 10.10. Let A be a commutative C^*-algebra with identity I, and let Δ denote its structure space. Verify the following properties of the Gelfand transform on A.

(a) x is invertible if and only if $\hat{x}$ is never 0.

(b) $x = yy^*$ if and only if $\hat{x} \geq 0$.

(c) x is a unitary element of A if and only if $|\hat{x}| \equiv 1$.

(d) A contains a nontrivial projection if and only if Δ is not connected.

EXERCISE 10.11. Let A and B be commutative C^*-algebras, each having an identity, and let Δ_A and Δ_B denote their respective structure spaces. Suppose T is a (not a priori continuous) homomorphism of the algebra A into the algebra B, and define $T' : \Delta_B \to \Delta_A$ by

$$T'(\phi) = \phi \circ TT.$$

(a) Prove that T' is continuous from Δ_B into Δ_A.

(b) Show that $\hat{x}(T'(\phi)) = \widehat{T(x)}(\phi)$ for each $x \in A$.

(c) Show that $\|T(x)\| \leq \|x\|$ and conclude that T is necessarily continuous.

(d) Prove that T' is onto if and only if T is 1-1. HINT: T is not 1-1 if and only if there exists a nontrivial continuous function on Δ_A that is identically 0 on the range of T'.

(e) Prove that T' is 1-1 if and only if T is onto.

(f) Prove that T' is a homeomorphism of Δ_B onto Δ_A if and only if T is an isomorphism of A onto B.

(g) Verify that parts a and b hold for any pair of Banach algebras A and B.

EXERCISE 10.12. (Independence of the Spectrum)

(a) Suppose B is a commutative C^*-algebra and that A is a sub-C^*-algebra of B. Let x be an element of A. Prove that $\mathrm{sp}_A(x) = \mathrm{sp}_B(x)$. HINT: Let θ be the injection map of A into B.

(b) Suppose C is a (not necessarily commutative) C^*-algebra, and let x be a normal element of C. Suppose A is the smallest sub-C^*-algebra of C that contains x x^*, and I. Prove that $\mathrm{sp}_A(x) = \mathrm{sp}_C(x)$. HINT: If $\lambda \in \mathrm{sp}_A(x)$, and $\lambda I - x$ has an inverse in C, let B be the smallest sub-C^*-algebra of C containing x, I, and $(\lambda I - x)^{-1}$. Then use part a.

(c) Let H be a separable Hilbert space, and let T be a normal element of $B(H)$. Let A be the smallest sub-C^*-algebra of $B(H)$ containing T, T^*, and I. Show that the spectrum $\mathrm{sp}(T)$ of the operator T coincides with the spectrum $\mathrm{sp}_A(T)$ of T thought of as an element of A.

THEOREM 10.9. (Spectral Theorem) *Let H be a separable Hilbert space, let A be a separable, commutative, sub-C^*-algebra of $B(H)$ that contains the identity operator I, and let Δ denote the structure space of A. Write $\mathcal{B}$ for the σ-algebra of Borel subsets of Δ. Then there exists a unique H-projection-valued measure p on $(\Delta, \mathcal{B})$ such that for every operator $S \in A$ we have*

$$S = \int \hat{S}\, dp.$$

That is, the inverse of the Gelfand transform is the integral with respect to p.

PROOF. Since A contains I, we know that Δ is compact and metrizable. Since the inverse T of the Gelfand transform is an isometric isomorphism of the Banach algebra $C(\Delta)$ onto A, we see that T satisfies the three conditions of Theorem 9.7.

(1) $T(fg) = T(f)T(g)$ for all $f, g \in C(\Delta)$.
(2) $T(\overline{f}) = [T(f)]^*$ for all $f \in C(\Delta)$.
(3) $T(1) = I$.

The present theorem then follows immediately from Theorem 9.7.

THEOREM 10.10. (Spectral Theorem for a Bounded Normal Operator) *Let T be a bounded normal operator on a separable Hilbert space H. Then there exists a unique H-projection-valued measure p on $(\mathbb{C}, \mathcal{B})$ such that*

$$T = \int f\, dp = \int f(\lambda)\, dp(\lambda),$$

where $f(\lambda) = \lambda$. We also use the notation

$$T = \int \lambda\, dp(\lambda).$$

Furthermore, $p_{\mathrm{sp}(T)} = I$; i.e., p is supported on the spectrum of T.

PROOF. Let A_0 be the set of all elements $S \in B(H)$ of the form

$$S = \sum_{i=0}^{n} \sum_{j=0}^{m} a_{ij} T^i T^{*j},$$

where each $a_{ij} \in \mathbb{C}$, and let A be the closure in $B(H)$ of A_0. We have that A is the smallest sub-C^*-algebra of $B(H)$ that contains T, T^*, and I. It follows that A is a separable commutative sub-C^*-algebra of $B(H)$ that contains I. If Δ denotes the structure space of A, then, by Theorem 10.9, there exists a unique projection-valued measure q on $(\Delta, \mathcal{B})$ such that

$$S = \int \hat{S} \, dq = \int \hat{S}(\phi) \, dq(\phi)$$

for every $S \in A$.

Note next that the function $\hat{T}$ is 1-1 on Δ. For, if $\hat{T}(\phi_1) = \hat{T}(\phi_2)$, then $\widehat{T^*}(\phi_1) = \widehat{T^*}(\phi_2)$, and hence $\hat{S}(\phi_1) = \hat{S}(\phi_2)$ for every $S \in A_0$. Therefore, $\hat{S}(\phi_1) = \hat{S}(\phi_2)$ for every $S \in A$, showing that $\phi_1 = \phi_2$. Hence, $\hat{T}$ is a homeomorphism of Δ onto the subset $\mathrm{sp}_A(T)$ of $\mathbb{C}$. By part c of Exercise 10.12, $\mathrm{sp}_A(T) = \mathrm{sp}(T)$.

Define a projection-valued measure $p = \hat{D}_* q$ on $\mathrm{sp}(T)$ by

$$p_E = \hat{T}_* q_E = q_{\hat{T}^{-1}(E)}.$$

See part c of Exercise 9.3. Then p is a projection-valued measure on $(\mathbb{C}, \mathcal{B})$, and p is supported on $\mathrm{sp}(T)$.

Now, let f be the identity function on $\mathbb{C}$, i.e., $f(\lambda) = \lambda$. Then, by Exercise 9.13, we have that

$$\begin{aligned}
\int \lambda \, dp(\lambda) &= \int f \, dp \\
&= \int (f \circ \hat{T}) \, dq \\
&= \int \hat{T} \, dq \\
&= T,
\end{aligned}$$

as desired.

Finally, let us show that the projection-valued measure p is unique. Suppose p' is another projection-valued measure on $(\mathbb{C}, \mathcal{B})$, supported on $\mathrm{sp}(T)$, such that

$$T = \int \lambda \, dp'(\lambda) = \int \lambda \, dp(\lambda).$$

It follows also that

$$T^* = \int \bar{\lambda} \, dp'(\lambda) = \int \bar{\lambda} \, dp(\lambda).$$

Then, for every function P of the form

$$P(\lambda) = \sum_{i=1}^{n} \sum_{j=1}^{m} c_{ij} \lambda^i \bar{\lambda}^j,$$

we have

$$\int P(\lambda) \, dp'(\lambda) = \int P(\lambda) \, dp(\lambda).$$

Whence, by the Stone-Weierstrass Theorem,

$$\int f(\lambda) \, dp'(\lambda) = \int f(\lambda) \, dp(\lambda)$$

for every continuous complex-valued function f on $\mathrm{sp}(T)$. If $q' = \hat{T}_*^{-1} p'$ is the projection-valued measure on Δ defined by

$$q'_E = p'_{\hat{T}(E)},$$

then, for any continuous function g on Δ, we have

$$\int g \, dq' = \int (g \circ \hat{T}^{-1}) \, dp'$$

$$= \int (g \circ \hat{T}^{-1}) \, dp$$

$$= \int (g \circ \hat{T}^{-1} \circ \hat{T}) \, dp$$

$$= \int g \, dp.$$

So, by the uniqueness assertion in the general spectral theorem, we have that $q' = q$. But then

$$p' = \hat{T}_* q' = \hat{T}_* q = p,$$

and the uniqueness is proved.

DEFINITION. The projection-valued measure p, associated as in the above theorem to a normal operator T, is called the *spectral measure* for T.

The next result is an immediate consequence of the preceding theorem.

THEOREM 10.11. (Spectral Theorem for a Bounded Selfadjoint Operator) *Let H be a separable Hilbert space, and let T be a selfadjoint element in $B(H)$. Then there exists a unique projection-valued measure p on $(\mathbb{R}, \mathcal{B})$ for which $T = \int \lambda\, dp(\lambda)$. Further, p is supported on the spectrum of T.*

REMARK. A slightly different notation is frequently used to indicate the spectral measure for a selfadjoint operator. Instead of writing $T = \int \lambda\, dp(\lambda)$, one often writes $T = \int \lambda\, dE_\lambda$. Also, such a projection-valued measure is sometimes referred to as a *resolution of the identity.*

EXERCISE 10.13. Let T be a normal operator in $B(H)$ and let p be its spectral measure.

(a) If U is a nonempty (relatively) open subset of $\mathrm{sp}(T)$, show that $p_U \neq 0$. If U is an infinite set, show that the range of p_U is infinite dimensional.

(b) Show that if E is a closed subset of $\mathbb{C}$ for which $p_E = I$, then E contains $\mathrm{sp}(T)$. Conclude that the smallest closed subset of $\mathbb{C}$ that supports p is the spectrum of T.

(c) If T is invertible, show that the function $1/\lambda$ is bounded on $\mathrm{sp}(T)$ and that $T^{-1} = \int (1/\lambda)\, dp(\lambda)$.

(d) If $\mathrm{sp}(T)$ contains at least two distinct points, show that $T = T_1 + T_2$, where T_1 and T_2 are both nonzero normal operators and $T_1 \circ T_2 = 0$.

(e) Suppose S is a bounded operator on H that commutes with T. Prove that S commutes with every projection p_E for E a Borel subset of $\mathrm{sp}(T)$. HINT: Do this first for open subsets of $\mathrm{sp}(T)$, and then consider the collection of all sets E for which $p_E S = S p_E$. (It is a monotone class.)

EXERCISE 10.14. Let T be a normal operator on a separable Hilbert space H, let A be a sub-C^*-algebra of $B(H)$ that contains T and I, let f be a continuous complex-valued function on the spectrum $\mathrm{sp}(T)$ of T, and suppose S is an element of A for which $\hat{S} = f \circ \hat{T}$.

(a) Show that the spectrum $\mathrm{sp}(S)$ of S equals $f(\mathrm{sp}(T))$. Compare this result with the spectral mapping theorem (part e of Exercise 10.7).

(b) Let p^T denote the spectral measure for T and p^S denote the spectral measure for S. In the notation of Exercises 9.3 and 9.13, show that

$$p^S = f_*(p^T).$$

HINT: Show that $S = \int \lambda\, df_*(p^T)(\lambda)$, and then use the uniqueness assertion in the Spectral Theorem for a normal operator.

(c) Apply parts a and b to describe the spectral measures for $S = q(T)$ for q a polynomial and $S = e^T$.

EXERCISE 10.15. Let p be an H-projection-valued measure on the Borel space $(S, \mathcal{B})$. If f is an element of $L^\infty(p)$, define the *essential range* of f to be the set of all $\lambda \in \mathbb{C}$ for which

$$p_{f^{-1}(B_\epsilon(\lambda))} \neq 0$$

for every $\epsilon > 0$.

(a) Let f be an element of $L^\infty(p)$. If T is the bounded normal operator $\int f \, dp$, show that the spectrum of T coincides with the essential range of f. See part e of Exercise 9.10.

(b) Let f be an element of $L^\infty(p)$, and let $T = \int f \, dp$. Prove that the spectral measure q for T is the projection-valued measure $f_* p$. See Exercises 9.3 and 9.13.

EXERCISE 10.16. Let (S, μ) be a σ-finite measure space. For each $f \in L^\infty(\mu)$, let m_f denote the multiplication operator on $L^2(\mu)$ given by $m_f g = fg$. Let p denote the canonical projection-valued measure on $L^2(\mu)$.

(a) Prove that the operator m_f is a normal operator and that

$$m_f = \int f \, dp.$$

Find the spectrum $\mathrm{sp}(m_f)$ of m_f.

(b) Using $S = [0, 1]$ and μ as Lebesgue measure, find the spectrum and spectral measures for the following m_f's:

(1) $f = \chi_{[0,1/2]}$,
(2) $f(x) = x$,
(3) $f(x) = x^2$,
(4) $f(x) = \sin(2\pi x)$, and
(5) f is a step function $f = \sum_{i=1}^n a_i \chi_{I_i}$, where the a_i's are complex numbers and the I_i's are disjoint intervals.

(c) Let S and μ be as in part b. Compute the spectrum and spectral measure for m_f if f is the Cantor function.

DEFINITION. We say that an operator $T \in B(H)$ is *diagonalizable* if it can be represented as the integral of a function with respect to a projection-valued measure. That is, if there exists a Borel space $(S, \mathcal{B})$ and an H-projection-valued measure p on $(S, \mathcal{B})$ such that $T = \int f \, dp$

for some bounded B-measurable function f. A collection B of operators is called *simultaneously diagonalizable* if there exists a projection-valued measure p on a Borel space (S, B) such that each element of B can be represented as the integral of a function with respect to p.

REMARK. Theorem 10.11 and Theorem 10.10 show that selfadjoint and normal operators are diagonalizable. It is also clear that simultaneously diagonalizable operators commute.

EXERCISE 10.17. (a) Let H be a separable Hilbert space. Suppose B is a commuting, separable, selfadjoint subset of $B(H)$. Prove that the elements of B are simultaneously diagonalizable.

(b) Let H be a separable Hilbert space. Show that a separable, selfadjoint collection S of operators in $B(H)$ is simultaneously diagonalizable if and only if S is contained in a commutative sub-C^*-algebra of $B(H)$.

(c) Let A be an $n \times n$ complex matrix for which $a_{ij} = \overline{a_{ji}}$. Use the Spectral Theorem to show that there exists a unitary matrix U such that UAU^{-1} is diagonal. That is, use the Spectral Theorem to prove that every Hermitian matrix can be diagonalized.

One of the important consequences of the spectral theorem is the following:

THEOREM 10.12. (Stone's Theorem) *Let $t \to U_t$ be a map of $\mathbb{R}$ into the set of unitary operators on a separable Hilbert space H, and suppose that this map satisfies:*

(1) $U_{t+s} = U_t \circ U_s$ *for all* $t, s \in \mathbb{R}$.
(2) *The map* $t \to (U_t(x), y)$ *is continuous for every pair* $x, y \in H$.

Then there exists a unique projection-valued measure p on $(\mathbb{R}, B)$ such that

$$U_t = \int e^{-2\pi i \lambda t}\, dp(\lambda)$$

for each $t \in \mathbb{R}$.

PROOF. For each $f \in L^1(\mathbb{R})$, define a map L_f from $H \times H$ into $\mathbb{C}$ by

$$L_f(x, y) = \int_{\mathbb{R}} f(s)((U_s(x)), y)\, ds.$$

It follows from Theorem 8.5 (see the exercise below) that for each $f \in L^1(\mathbb{R})$ there exists a unique element $T_f \in B(H)$ such that

$$L_f(x, y) = (T_f(x), y)$$

for all $x, y \in H$. Let B denote the set of all operators on H of the form T_f for $f \in L^1(\mathbb{R})$. Again by the exercise below, it follows that B is a separable commutative selfadjoint subalgebra of $B(H)$.

We claim first that the subspace H_0 spanned by the vectors of the form $y = T_f(x)$, for $f \in L^1(\mathbb{R})$ and $x \in H$, is dense in H. Indeed, if $z \in H$ is orthogonal to every element of H_0, then

$$0 = (T_f(z), z)$$
$$= \int_{\mathbb{R}} f(s)(U_s(z), z)\, ds$$

for all $f \in L^1(\mathbb{R})$, whence

$$(U_s(z), z) = 0$$

for almost all $s \in \mathbb{R}$. But, since this is a continuous function of s, it follows that

$$(U_s(z), z) = 0$$

for all s. In particular,

$$(z, z) = (U_0(z), z) = 0,$$

proving that H_0 is dense in H as claimed.

We let A denote the smallest sub-C^*-algebra of $B(H)$ that contains B and the identity operator I, and we denote by Δ the structure space of A. We see that A is the closure in $B(H)$ of the set of all elements of the form $\lambda I + T_f$, for $\lambda \in \mathbb{C}$ and $f \in L^1\mathbb{R})$. So A is a separable commutative C^*-algebra. Again, by Exercise 10.18 below, we have that the map T that sends $f \in L^1(\mathbb{R})$ to the operator T_f is a norm-decreasing homomorphism of the Banach $*$-algebra $L^1(\mathbb{R})$ into the C^*-algebra A. Recall from Exercise 10.5 that the structure space of the Banach algebra $L^1(\mathbb{R})$ is identified, specifically as in that exercise, with the real line $\mathbb{R}$. With this identification, we define $T' : \Delta \to \mathbb{R}$ by

$$T'(\phi) = \phi \circ T.$$

Because the topologies on the structures spaces of A and $L^1(\mathbb{R})$ are the weak* topologies, it follows directly that T' is continuous. For each $f \in L^1(\mathbb{R})$ we have the formula

$$\hat{f}(T'(\phi)) = [T'(\phi)](f) = \phi(T_f) = \widehat{T_f}(\phi).$$

By the general Spectral Theorem, we let q be the unique projection-valued measure on Δ for which

$$S = \int \hat{S}(\phi) \, dq(\phi)$$

for all $S \in A$, and we set $p = T'_* q$. Then p is a projection-valued measure on $(\mathbb{R}, \mathcal{B})$, and we have

$$\int \hat{f} \, dp = \int (\hat{f} \circ T') \, dq$$
$$= \int \hat{f}(T'(\phi)) \, dq(\phi)$$
$$= \int \widehat{T_f}(\phi) \, dq(\phi)$$
$$= T_f$$

for all $f \in L^1(\mathbb{R})$.

Now, for each $f \in L^1(\mathbb{R})$ and each real t we have

$$(U_t(T_f(x)), y) = \int_{\mathbb{R}} f(s)(U_t(U_s(x)), y) \, ds$$
$$= \int_{\mathbb{R}} f(s)(U_{t+s}(x), y) \, ds$$
$$= \int_{\mathbb{R}} f_{-t}(s)(U_s(x), y) \, ds$$
$$= (T_{f_{-t}}(x), y)$$
$$= ([\int \widehat{f_{-t}}(\lambda) \, dp(\lambda)](x), y)$$
$$= ([\int e^{-2\pi i \lambda t} \hat{f}(\lambda) \, dp(\lambda)](x), y)$$
$$= ([\int e^{-2\pi i \lambda t} \, dp(\lambda)](T_f(x)), y),$$

where f_{-t} is defined by $f_{-t}(x) = f(x - t)$. So, because the set H_0 of all vectors of the form $T_f(x)$ span a dense subspace of H,

$$U_t = \int e^{-2\pi i \lambda t} \, dp(\lambda),$$

as desired.

We have left to prove the uniqueness of p. Suppose $\tilde{p}$ is a projection-valued measure on $(\mathbb{R}, \mathcal{B})$ for which $U_t = \int e^{-2\pi i \lambda t} \, d\tilde{p}(\lambda)$ for all t. Now for each vector $x \in H$, define the two measures μ_x and $\tilde{\mu}_x$ by

$$\mu_x(E) = (p_E(x), x)$$

and

$$\tilde{\mu}_x(E) = (\tilde{p}_E(x), x).$$

Our assumption on $\tilde{p}$ implies then that

$$\int e^{-2\pi i \lambda t} \, d\mu_x(\lambda) = \int e^{-2\pi i \lambda t} \, d\tilde{\mu}_x(\lambda)$$

for all real t. Using Fubini's theorem we then have for every $f \in L^1(\mathbb{R})$ that

$$\int \hat{f}(\lambda) \, d\mu_x(\lambda) = \int \int f(t) e^{-2\pi i \lambda t} \, dt \, d\mu_x(\lambda)$$

$$= \int f(t) \int e^{-2\pi i \lambda t} \, d\mu_x(\lambda) \, dt$$

$$= \int f(t) \int e^{-2\pi i \lambda t} \, d\tilde{\mu}_x(\lambda) \, dt$$

$$= \int \hat{f}(\lambda) \, d\tilde{\mu}_x(\lambda).$$

Since the set of Fourier transforms of L^1 functions is dense in $C_0(\mathbb{R})$, it then follows that

$$\int g \, d\mu_x = \int g \, d\tilde{\mu}_x$$

for every $g \in C_0(\mathbb{R})$. Therefore, by the Riesz representation theorem, $\mu_x = \tilde{\mu}_x$. Consequently, $p = \tilde{p}$ (see part d of Exercise 9.2), and the proof is complete.

EXERCISE 10.18. Let the map $t \to U_t$ be as in the theorem above.

(a) Prove that U_0 is the identity operator on H and that $U_t^* = U_{-t}$ for all t.

(b) If $f \in L^1(\mathbb{R})$, show that there exists a unique element $T_f \in B(H)$ such that

$$\int_{\mathbb{R}} f(s)(U_s(x), y) \, ds = (T_f(x), y)$$

for all $x, y \in H$. HINT: Use Theorem 8.5.

(c) Prove that the assignment $f \to T_f$ defined in part b satisfies

$$T_{f*g} = T_f \circ T_g$$

for all $f, g \in L^1(\mathbb{R})$ and

$$T_{f^*} = T_f^*$$

for all $f \in L^1(\mathbb{R})$, where

$$f^*(s) = \overline{f(-s)}.$$

(d) Conclude that the set of all T_f's, for $f \in L^1(\mathbb{R})$, is a separable commutative selfadjoint algebra of operators.

EXERCISE 10.19. Let H be a separable Hilbert space, let A be a separable, commutative, sub-C^*-algebra of $B(H)$, assume that A contains the identity operator I, and let Δ denote the structure space of A. Let x be a vector in H, and let M be the closure of the set of all vectors $T(x)$, for $T \in A$. That is, M is a *cyclic subspace* for A. Prove that there exists a finite Borel measure μ on Δ and a unitary operator U of $L^2(\mu)$ onto M such that

$$U^{-1} \circ T \circ U = m_{\hat{T}}$$

for every $T \in A$. HINT: Let G denote the inverse of the Gelfand transform of A. Define a positive linear functional I on $C(\Delta)$ by $I(f) = ([G(f)](x), x)$, use the Riesz Representation Theorem to get a measure μ, and then define $U(f) = [G(f)](x)$ on the dense subspace $C(\Delta)$ of $L^2(\mu)$.

CHAPTER XI

APPLICATIONS OF SPECTRAL THEORY

Let H be a separable, infinite-dimensional, complex Hilbert space. We exploit properties of the Spectral Theorem to investigate and classify operators on H. As usual, all Hilbert spaces considered will be assumed to be complex and separable, even if it is not explicitly stated.

If T is an element of the C^*-algebra $B(H)$, recall that the *resolvent* of T is the set $\mathrm{res}(T)$ of all complex numbers λ for which $\lambda I - T$ has a two-sided inverse in $B(H)$. The *spectrum* $\mathrm{sp}(T)$ of T is the complement of the resolvent of T. That is, λ belongs to $\mathrm{sp}(T)$ if $\lambda I - T$ does not have a bounded two-sided inverse.

THEOREM 11.1. (Existence of Positive Square Roots of Positive Operators) *Let H be a Hilbert space, and let T be a positive operator in $B(H)$; i.e., $(T(x), x) \geq 0$ for all $x \in H$. Then:*

(1) *There exists an element R in $B(H)$ such that $T = R^* R$.*
(2) *There exists a unique positive square root of T, i.e., a unique positive operator S such that $T = S^2$.*
(3) *If T is invertible, then its positive square root S is also invertible.*

PROOF. We know that a positive operator T is necessarily selfadjoint. Hence, writing $T = \int_{\mathbb{R}} \lambda \, dp(\lambda)$, let us show that $p_{(-\infty,0)} = 0$. That is, the spectrum of T is contained in the set of nonnegative real numbers. If not, there must exist a $\delta > 0$ such that $p_{(-\infty,-\delta]} \neq 0$. If x is a nonzero

211

vector in the range of $p_{(-\infty,-\delta]}$, then

$$(T(x), x) = (T(p_{(-\infty,-\delta]}(x)), x)$$
$$= \int_{-\infty}^{-\delta} \lambda \, d\mu_x(\lambda)$$
$$\leq -\delta \|x\|^2$$
$$< 0.$$

But this would imply that T is not a positive operator. Hence, p is supported on $[0, \infty)$. Clearly then $T = S^2 = S^* S$, where

$$S = \int \sqrt{\lambda} \, dp(\lambda).$$

Setting $R = S$ gives part 1.

Since S is the integral of a nonnegative function with respect to a projection-valued measure, it follows that S is a positive operator, so that S is a positive square root of T. We know from the Weierstrass Theorem that the continuous function $\sqrt{\lambda}$ is the uniform limit on the compact set $\mathrm{sp}(T)$ of a sequence of polynomials in λ. It follows that S is an element of the smallest sub-C^*-algebra A of $B(H)$ containing T and I.

Now, if S' is any positive square root of T, then S' certainly commutes with $T = S'^2$. Hence, S' commutes with every element of the algebra A and hence in particular with S. Let A' be the smallest sub-C^*-algebra of $B(H)$ that contains I, T and S'. Then A' is a separable commutative C^*-algebra with identity, and S and S' are two positive elements of A' whose square is T. But the Gelfand transform on A' is 1-1 and, by part 1 of this theorem and part b of Exercise 10.10, sends both S and S' to the function $\sqrt{\hat{T}}$. Hence, $S = S'$, completing the proof of part 2.

Finally, if T is invertible, say $TU = I$, then $S(SU) = I$, showing that S has a right inverse. Also, $(US)S = I$, showing that S also has a left inverse so is invertible.

EXERCISE 11.1. (a) Let T be a selfadjoint element of $B(H)$. Prove that there exist unique positive elements T_+ and T_- such that $T = T_+ - T_-$, T_+ and T_- commute with T and with each other, and $T_+ T_- = 0$. HINT: Use the Gelfand transform. T_+ and T_- are called the *positive* and *negative* parts of the selfadjoint operator T.

(b) Let T, T_+, and T_- be as in part a. Verify that, for any $x \in H$, we have

$$\|T(x)\|^2 = \|T_+(x)\|^2 + \|T_-(x)\|^2.$$

Show further that $\sqrt{T^2} = T_+ + T_-$. How are T_+ and T_- represented in terms of the spectral measure for T? Conclude that every element $T \in B(H)$ is a complex linear combination of four positive operators.

(c) Suppose T is a positive operator, and let p denote its spectral measure. Suppose $0 \le a < b < \infty$ and that x is an element of the range of $p_{[a,b]}$. Show that $a\|x\| \le \|T(x)\| \le b\|x\|$.

(d) If U is a unitary operator, prove that there exists a selfadjoint operator $T \in B(H)$ for which $U = e^{iT}$. HINT: Show that the function $\hat{U} = e^{ir(x)}$ for some bounded real-valued Borel function r.

(e) If T is a positive operator, show that $I + T$ is invertible.

(f) Suppose T and S are invertible positive operators that commute. Assume that $S - T$ is a positive operator, i.e., that $S \ge T$. Prove that $T^{-1} - S^{-1}$ is a positive operator, i.e., that $T^{-1} \ge S^{-1}$.

(g) Suppose T is a positive operator and that S is a positive invertible operator not necessarily commuting with T. Prove that $S + T$ is positive and invertible.

DEFINITION. Let M be a subspace of a Hilbert space H. By a *partial isometry* of M into H we mean an element V of $B(H)$ that is an isometry on M and is 0 on the orthogonal complement $M^\perp$ of M.

EXERCISE 11.2. Let V be a partial isometry of M into H.

(a) Show that $(V(x), V(y)) = (x, y)$ for all $x, y \in M$.

(b) Show that V^*V is the projection p_M of H onto M and that VV^* is the projection $p_{\overline{V(M)}}$ of H onto $\overline{V(M)}$.

(c) Show that V^* is a partial isometry of $V(M)$ into H.

(d) Let H be the set of square-summable sequences $\{a_1, a_2, \ldots\}$, and let M be the subspace determined by the condition $a_1 = 0$. Define $V : M \to H$ by $[V(\{a_n\})]_n = a_{n+1}$. Show that V is a partial isometry of M into H. Compute V^*. (This V is often called the *unilateral shift*.)

THEOREM 11.2. (Polar Decomposition Theorem) *Let H be a Hilbert space, and let T be an element of $B(H)$. Then there exist unique operators P and V satisfying:*

(1) *P is a positive operator, and V is a partial isometry from the range of P into H.*

(2) *$T = VP$ and $P = V^*T$.*

Moreover, if T is invertible, then P is invertible and V is a unitary operator.

PROOF. Let $P = \sqrt{T^*T}$. Then P is positive. Observe that $\|P(x)\| = \|T(x)\|$ for all x, whence, if $P(x) = 0$ then $T(x) = 0$. Indeed,

$$(P(x), P(x)) = (P^2(x), x) = (T^*T(x), x) = (T(x), T(x)).$$

Therefore, the map V, that sends $P(x)$ to $T(x)$, is an isometry from the range M of P onto the range of T. Defining V to be its unique isometric extension to $\bar{M}$ on all of $\bar{M}$ and to be 0 on the orthogonal complement $M^\perp$ of M, we have that V is a partial isometry of M into H. Further, $T(x) = V(P(x))$, and $T = VP$, as desired. Further, from the preceding exercise, V^*V is the projection onto the range M of P, so that $V^*T = V^*VP = P$.

If Q is a positive operator and W is a partial isometry of the range of Q into H for which $T = WQ$ and $Q = W^*T$, then W^*W is the projection onto the range of Q. Hence,

$$T^*T = QW^*WQ = Q^2,$$

whence $Q = P$ since positive square roots are unique. But then $V = W$, since they are both partial isometries of the range of P into H, and they agree on the range of P. Therefore, the uniqueness assertion of the theorem is proved.

Finally, if T is invertible, then P is invertible, and the partial isometry $V = TP^{-1}$ is invertible. An isometry that is invertible is of course a unitary operator.

DEFINITION. The operator $P = \sqrt{T^*T}$ of the preceding theorem is called the *absolute value* of T and is often denoted by $|T|$.

REMARK. We have defined the absolute value of an operator T to be the square root of the positive operator T^*T. We might well have chosen to define the absolute value of T to be the square root of the (probably different) positive operator TT^*. Though different, either of these choices would have sufficed for our eventual purposes. See part c of the following exercise.

EXERCISE 11.3. Let T be an operator in $B(H)$.
(a) Prove that $\||T|(x)\| = \|T(x)\|$ for every $x \in H$.
(b) If T is a selfadjoint operator, and we write $T = T_+ - T_-$ (as in Exercise 11.1), show that $|T| = T_+ + T_-$.

(c) Show that there exists a unique positive operator P' and a unique partial isometry V' of the range of T^* into H such that $T = P'V'$ and $P' = TV'^*$. Are P' and V' identical with the P and V of the preceding theorem?

We introduce next a number of definitions concerning the spectrum of an operator.

DEFINITION. Let T be a normal operator, and let p be its spectral measure.

(1) A complex number λ is said to belong to the *point spectrum* $\mathrm{sp}_p(T)$ of T if $p_{\{\lambda\}} \neq 0$. In this case we say that the *multiplicity* of λ is the dimension $m(\lambda)$ of the range of $p_{\{\lambda\}}$.

(2) An element λ of the spectrum of T, which is not in the point spectrum, is said to belong to the *continuous spectrum* $\mathrm{sp}_c(T)$ of T. The *multiplicity* $m(\lambda)$ of an element λ of the continuous spectrum is defined to be 0.

(3) A complex number λ is said to belong to the *discrete spectrum* $\mathrm{sp}_d(T)$ of T if $\{\lambda\}$ is an isolated point in the compact set $\mathrm{sp}(T)$. Note that if $\lambda \in \mathrm{sp}_d(T)$, then $\{\lambda\}$ is a relatively open subset of $\mathrm{sp}(T)$. It follows then from part a of Exercise 10.13 that $\mathrm{sp}_d(T) \subseteq \mathrm{sp}_p(T)$.

(4) A complex number λ is said to belong to the *essential spectrum* $\mathrm{sp}_e(T)$ if it is not an element of the discrete spectrum with finite multiplicity.

(5) T is said to have *purely atomic spectrum* if p is supported on a countable subset of $\mathbb{C}$.

EXERCISE 11.4. (Characterization of the Point Spectrum) Suppose T is a normal operator, that p is its spectral measure, and that v is a nonzero vector for which $T(v) = 0$. Write μ_v for the measure on $\mathrm{sp}(T)$ given by $\mu_v(E) = (p_E(v), v)$.
(a) Prove that $0 \in \mathrm{sp}(T)$.
(b) Show that $\int \lambda^n \, d\mu_v(\lambda) = 0$ for all positive integers n.
(c) Prove that $\int f(\lambda) \, d\mu_v(\lambda) = f(0)$ for all $f \in C(\mathrm{sp}(T))$.
(d) Show that $\mu_v = \delta_0$, whence $p_{\{0\}} \neq 0$.
(e) Let T be an arbitrary normal operator. Prove that $\lambda_0 \in \mathrm{sp}_p(T)$ if and only if λ_0 is an eigenvalue for T. HINT: Write $S = T - \lambda_0 I$, and use Exercise 10.14.

EXERCISE 11.5. Let H be the Hilbert space l^2 consisting of the square summable sequences $\{a_1, a_2, \dots\}$. Let $r_1, r_2, \dots$ be a sequence of (not necessarily distinct) numbers in the interval $[0,1]$, and define an operator T on l^2 by

$$T(\{a_n\}) = \{r_n a_n\}.$$

(a) Prove that T is a selfadjoint operator–even a positive operator.

(b) Show that the point spectrum of T is the set of r_n's.

(c) Find the spectrum of T.

(d) Find the discrete spectrum of T.

(e) Find the essential spectrum of T.

(f) Choose the sequence $\{r_n\}$ so that $\mathrm{sp}_d(T) \subset \mathrm{sp}_p(T)$ and $\mathrm{sp}_e(T) \subset \mathrm{sp}(T)$.

(g) Construct a sequence $\{T_j\}$ of positive operators that converges in norm to a positive operator T, but for which the sequence $\{\mathrm{sp}_d(T_j)\}$ of subsets of $\mathbb{R}$ in no way converges to $\mathrm{sp}_d(T)$. Test a few other conjectures concerning the continuity of the map $T \to \mathrm{sp}(T)$.

THEOREM 11.3. Let H be a separable Hilbert space, and let T be a normal operator in $B(H)$. Then the following are equivalent:

(1) T has purely atomic spectrum.

(2) There exists an orthonormal basis for H consisting of eigenvectors for T.

(3) There exists a sequence $\{p_i\}$ of pairwise orthogonal projections and a sequence $\{\lambda_i\}$ of complex numbers such that

$$I = \sum_{i=1}^{\infty} p_i$$

and

$$T = \sum_{i=1}^{\infty} \lambda_i p_i.$$

PROOF. If T has purely atomic spectrum, and if $\lambda_1, \lambda_2, \ldots$ denotes a countable set on which the spectral measure p is concentrated, let $p_i = p_{\{\lambda_i\}}$. Then the p_i's are pairwise orthogonal, and

$$I = \sum_{i=1}^{\infty} p_i,$$

and

$$T = \int \lambda \, dp(\lambda)$$

$$= \sum_{i=1}^{\infty} \lambda_i p_{\{\lambda_i\}}$$

$$= \sum_{i=1}^{\infty} \lambda_i p_i,$$

showing that 1 implies 3.

Next, suppose $T = \sum_{i=1}^{\infty} \lambda_i p_i$, where $\{p_i\}$ is a sequence of pairwise orthogonal projections for which $I = \sum p_i$. We may make an orthonormal basis for H by taking the union of orthonormal bases for the ranges M_{p_i} of the p_i's. Clearly, each vector in this basis is an eigenvector for T, whence, 3 implies 2.

Finally, suppose there exists an orthonormal basis for H consisting of eigenvectors for T, and let $\{\lambda_1, \lambda_2, \dots\}$ be the set of distinct eigenvalues for T. Because T is a bounded operator, this set of λ_i's is a bounded subset of $\mathbb{C}$. For each $i = 1, 2, \dots$, let M_i be the eigenspace corresponding to the eigenvalue λ_i, and write p_i for the projection onto M_i. Then the p_i's are pairwise orthogonal, and $I = \sum p_i$.

Now, for each subset $E \subseteq \mathbb{C}$, define

$$p_E = \sum_{\lambda_i \in E} p_i.$$

Then $E \to p_E$ is a projection-valued measure supported on the compact set $\overline{\{\lambda_i\}}$, and we let S be the normal operator given by $S = \int \lambda \, dp(\lambda)$. If $v \in M_i$, then v belongs to the range of $p_{\{\lambda_i\}}$, whence $v = p_{\{\lambda_i\}}(v)$. It follows then that

$$T(v) = \lambda_i v$$

$$= \lambda_i p_{\{\lambda_i\}}(v)$$

$$= [\int \lambda \chi_{\{\lambda_i\}}(\lambda) \, dp(\lambda)](v)$$

$$= [\int \lambda \, dp(\lambda)]([\int \chi_{\{\lambda_i\}}(\lambda) \, dp(\lambda)](v))$$

$$= S(p_{\{\lambda_i\}}(v))$$

$$= S(v).$$

Since this holds for each i, we have that $T = S$, showing that 2 implies 1.

The next theorem describes a subtle but important distinction between the spectrum and the essential spectrum. However, the true essence of the essential spectrum is only evident in Theorem 11.9.

THEOREM 11.4. *Let T be a normal operator on a separable Hilbert space H. Then*

(1) *$\lambda_0 \in sp(T)$ if and only if there exists a sequence $\{v_n\}$ of unit vectors in H such that*

$$\lim \|T(v_n) - \lambda_0 v_n\| = 0.$$

(2) *$\lambda_0 \in sp_e(T)$ if and only if there exists an infinite sequence $\{v_n\}$ of orthonormal vectors for which*

$$\lim \|T(v_n) - \lambda_0 v_n\| = 0.$$

PROOF. (1) If λ_0 belongs to the point spectrum of T, then there exists a unit vector v (any unit vector in the range of $p_{\{\lambda_0\}}$) such that $T(v) - \lambda_0 v = 0$. Therefore, the constant sequence $v_n \equiv v$ satisfies

$$\lim \|T(v_n) - \lambda_0 v_n\| = 0.$$

(2) If λ_0 belongs to the point spectrum of T, and the multiplicity $m(\lambda_0)$ is infinity, then there exists an infinite orthonormal sequence $\{v_n\}$ in the range of $p_{\{\lambda_0\}}$ such that $T(v_n) - \lambda_0 v_n \equiv 0$.

(3) Suppose $\lambda_0 \in sp(T)$ but $\lambda_0 \notin sp_d(T)$. For each positive integer k, let $U_k = sp(T) \cap B_{1/k}(\lambda_0)$. Then each U_k is a nonempty open subset of $sp(T)$, whence $p_{U_k} \neq 0$ for all k. In fact, since λ_0 is not a discrete point in the spectrum of T, there exists an increasing sequence $\{k_n\}$ of positive integers such that $p_{F_n} \neq 0$ for every n, where $F_n = U_{k_n} - U_{k_{n+1}}$. (Why?) Choosing v_n to be a unit vector in the range of the projection p_{F_n}, we see that the sequence $\{v_n\}$ is infinite and orthonormal. Further,

we have

$$\|T(v_n) - \lambda_0 v_n\| = \|T(p_{F_n}(v_n)) - \lambda_0 p_{F_n}(v_n)\|$$

$$= \|[\int \lambda \chi_{F_n}(\lambda)\, dp(\lambda)](v_n) - [\int \lambda_0 \chi_{F_n}(\lambda)\, dp(\lambda)](v_n)\|$$

$$= \|[\int (\lambda - \lambda_0)\chi_{F_n}(\lambda)\, dp(\lambda)](v_n)\|$$

$$\leq \| \int (\lambda - \lambda_0)\chi_{F_n}(\lambda)\, dp(\lambda)\|\|v_n\|$$

$$\leq \| \int (\lambda - \lambda_0)\chi_{B_{1/k_n}(\lambda_0)}(\lambda)\, dp(\lambda)\|$$

$$\leq 1/k_n.$$

This shows that $\lim_n \|T(v_n) - \lambda_0 v_n\| = 0$.

(4) If $\lambda_0 \notin \operatorname{sp}(T)$, let ϵ be a positive number such that $|\lambda_0 - \lambda| \geq \epsilon$ for every $\lambda \in \operatorname{sp}(T)$. Then the function $f(\lambda) = 1/(\lambda - \lambda_0)$ is bounded and continuous on $\operatorname{sp}(T)$, whence the operator $S = \int f(\lambda)\, dp(\lambda)$ is a bounded operator on H. Obviously, this operator S is $(T - \lambda_0 I)^{-1}$. So, if $\{v_n\}$ were a sequence of unit vectors, for which $\lim_n (T(v_n) - \lambda_0 v_n) = 0$, then $\lim v_n = \lim(T - \lambda_0 I)^{-1}((T - \lambda_0 I)(v_n)) = 0$, which is a contradiction.

The completion of this proof is left to the exercise that follows.

EXERCISE 11.6. Use results 1-4 above to complete the proof of Theorem 11.4.

We next introduce some important classes of operators on an infinite dimensional Hilbert space. Most of these classes are defined in terms of the spectral measures of their elements.

DEFINITION. Let H be an infinite-dimensional separable Hilbert space.

(1) An element $T \in B(H)$ is a *finite rank operator* if its range is finite dimensional.

(2) A positive operator T is a *compact operator* if it has purely atomic spectrum, and this spectrum consists of a (possibly finite) strictly decreasing sequence $\{\lambda_i\}$ of nonnegative numbers, such that $0 = \lim \lambda_i$, and such that the multiplicity $m(\lambda_i)$ is finite for every $\lambda_i > 0 \in \operatorname{sp}(T)$. (If the sequence $\lambda_1, \lambda_2, \ldots$ is finite, then the statement $0 = \lim \lambda_i$ means that $\lambda_N = 0$ for some (the last) N. Evidently each positive element λ_i of this spectrum is a discrete point, whence each positive λ_i of the spectrum is an eigenvalue for T.) A selfadjoint element $T = T_+ - T_- \in B(H)$

is a *compact operator* if its positive and negative parts T_+ and T_- are compact operators, and a general element $T = T_1 + iT_2 \in B(H)$ is a *compact operator* if its real and imaginary parts T_1 and T_2 are compact operators.

(3) A positive operator T is a *trace class operator* if it is a compact operator, with positive eigenvalues $\lambda_1, \lambda_2, \ldots$, for which

$$\sum \lambda_i m(\lambda_i) < \infty.$$

A selfadjoint element $T = T_+ - T_- \in B(H)$ is a *trace class operator* if its positive and negative parts T_+ and T_- are trace class operators, and a general $T = T_1 + iT_2 \in B(H)$ is a *trace class operator* if its real and imaginary parts T_1 and T_2 are trace class operators.

(4) A positive operator T is a *Hilbert-Schmidt operator* if it is a compact operator, with positive eigenvalues $\lambda_1, \lambda_2, \ldots$, for which

$$\sum \lambda_i^2 m(\lambda_i) < \infty.$$

A selfadjoint element $T = T_+ - T_- \in B(H)$ is a *Hilbert-Schmidt operator* if its positive and negative parts T_+ and T_- are Hilbert-Schmidt operators, and a general $T = T_1 + iT_2 \in B(H)$ is a *Hilbert-Schmidt operator* if its real and imaginary parts T_1 and T_2 are Hilbert-Schmidt operators.

EXERCISE 11.7. (a) Let T be in $B(H)$. Prove that the closure of the range of T is the orthogonal complement of the kernel of T^*. Conclude that T is a finite rank operator if and only if T^* is a finite rank operator.

(b) Show that the set of finite rank operators forms a two-sided self-adjoint ideal in $B(H)$.

(c) Show that T is a finite rank operator if and only if $|T|$ is a finite rank operator.

(d) Show that every finite rank operator is a trace class operator, and that every trace class operator is a Hilbert-Schmidt operator.

(e) Using multiplication operators on l^2 (see Exercise 11.5), show that the inclusions in part d are proper. Show also that the set of Hilbert-Schmidt operators is a proper subset of the set of compact operators on l^2 and that the set of compact operators is a proper subset of $B(l^2)$.

(f) Let T be a positive compact operator. Prove that there exists an orthonormal basis for H consisting of eigenvectors for T.

(g) Prove that every normal compact operator has purely atomic spectrum.

THEOREM 11.5. (Characterization of Compact Operators) *Suppose T is a bounded operator on a separable infinite-dimensional Hilbert space H. Then the following properties are equivalent:*

(1) *T is a compact operator.*

(2) *If $\{x_n\}$ is any bounded sequence of vectors in H, then $\{T(x_n)\}$ has a convergent subsequence.*

(3) *$T(B_1)$ has a compact closure in H.*

(4) *If $\{x_n\}$ is a sequence of vectors in H that converges weakly to 0, then the sequence $\{T(x_n)\}$ converges in norm to 0.*

(5) *T is the limit in $B(H)$ of a sequence of finite rank operators.*

PROOF. Let us first show that 1 implies 5. It will suffice to show this for T a positive compact operator. Thus, let $\{\lambda_1, \lambda_2, \ldots\}$ be the strictly decreasing (finite or infinite) sequence of positive elements of $\mathrm{sp}(T)$. Using the Spectral Theorem and the fact that T has purely atomic spectrum, write

$$T = \int \lambda \, dp(\lambda) = \sum_i \lambda_i p_{\{\lambda_i\}}.$$

Evidently, if there are only a finite number of λ_i's, then T is itself a finite rank operator, since the dimension of the range of each $p_{\{\lambda_i\}}$ is finite, and 5 follows. Hence, we may assume that the sequence $\{\lambda_i\}$ is infinite. Define a sequence $\{T_k\}$ of operators by

$$T_k = \sum_{i=1}^{k} \lambda_i p_{\{\lambda_i\}}$$

$$= \int \chi_{[\lambda_k, \infty)}(\lambda) \lambda \, dp(\lambda).$$

Then each T_k is a finite rank operator. Further,

$$\|T - T_k\| = \| \int \chi_{[0, \lambda_k)}(\lambda) \lambda \, dp(\lambda)\| \leq \lambda_k.$$

Hence, $T = \lim T_k$ in norm, giving 5.

We show next that 5 implies 4. Suppose then that $T = \lim T_k$ in norm, where each T_k is a finite rank operator. Let $\{x_n\}$ be a sequence in H that converges weakly to 0, and let $\epsilon > 0$ be given. Then, by the Uniform Boundedness Theorem, the sequence $\{x_n\}$ is uniformly bounded, and

$$\|T(x_n)\| \leq \|(T - T_k)x_n\| + \|T_k(x_n)\|.$$

Choose k so that $\|(T-T_k)(x_n)\| < \epsilon/2$ for all n. For this k, the sequence $\{T_k(x_n)\}$ is contained in the finite dimensional subspace M that is the range of T_k, and converges weakly to 0 there. Since all vector space topologies are identical on a finite dimensional space, we have that, for this fixed k, the sequence $\{T_k(x_n)\}$ also converges to 0 in norm. Choose N so that $\|T_k(x_n)\| < \epsilon/2$ for all $n \geq N$. Then $\|T(x_n)\| < \epsilon$ if $n \geq N$, and the sequence $\{T(x_n)\}$ converges to 0 in norm, as desired.

We leave to the exercises the fact that properties 2,3, and 4 are equivalent (for any element of $B(H)$). Let us show finally that 4 implies 1. Thus, suppose T satisfies 4. Then T^* also satisfies 4. For, if the sequence $\{x_n\}$ converges to 0 weakly, then the sequence $\{T^*(x_n)\}$ also converges to 0 weakly. Hence, the sequence $\{T(T^*(x_n))\}$ converges to 0 in norm. Since

$$\|T^*(x_n)\|^2 = (T^*(x_n), T^*(x_n)) = (T(T^*(x_n)), x_n) \leq \|T(T^*(x_n))\|\|x_n\|,$$

it follows that the sequence $\{T^*(x_n)\}$ converges to 0 in norm. Consequently, the real and imaginary parts T_1 and T_2 of T satisfy 4, and we may assume that T is selfadjoint. Write $T = T_+ - T_-$ in terms of its positive and negative parts. By part b of Exercise 11.1, we see that both T_+ and T_- satisfy 4, so that we may assume that T is a positive operator. Let p be the spectral measure for T, and note that for each positive ϵ, we must have that the range of $p_{(\epsilon,\infty)}$ must be finite dimensional. Otherwise, there would exist an orthonormal sequence $\{x_n\}$ in this range. Such an orthonormal sequence converges to 0 weakly, but, by part b of Exercise 9.11, $\|T(x_n)\| \geq \epsilon$ for all n, contradicting 4. Hence, $\mathrm{sp}(T) \cap (\epsilon, \infty)$ is a finite set for every positive ϵ, whence the spectrum of T consists of a decreasing sequence of nonnegative numbers whose limit is 0. It also follows as in the above that each $p_{\{\lambda\}}$, for $\lambda > 0 \in \mathrm{sp}(T)$, must have a finite dimensional range, whence T is a compact operator, completing the proof that 4 implies 1.

EXERCISE 11.8. (Completing the Proof of the Preceding Theorem) Let T be an arbitrary element of $B(H)$.

(a) Assume 2. Show that $T(B_1)$ is totally bounded in H, and then conclude that 3 holds. (A subset E of a metric space X is called *totally bounded* if for every positive ϵ the set E is contained in a finite union of sets of diameter less than ϵ.)

(b) Prove that 3 implies 4.

(c) Prove that 4 implies 2.

EXERCISE 11.9. (Properties of the Set of Compact Operators)

(a) Prove that the set K of all compact operators forms a proper closed two-sided selfadjoint ideal in the C^*-algebra $B(H)$.

(b) Prove that an element $T \in B(H)$ is a compact operator if and only if $|T|$ is a compact operator.

(c) Show that no compact operator can be invertible.

(d) Show that the essential spectrum of a normal compact operator is singleton 0.

THEOREM 11.6. (Characterization of Hilbert-Schmidt Operators) Let H be a separable infinite-dimensional Hilbert space.

(1) If T is any element of $B(H)$, then the extended real number

$$\sum_i \|T(\phi_i)\|^2$$

is independent of which orthonormal basis $\{\phi_i\}$ is used. Further,

$$\sum_i \|T(\phi_i)\|^2 = \sum_i \|T^*(\phi_i)\|^2.$$

(2) An operator T is a Hilbert-Schmidt operator if and only if

$$\sum_i \|T(\phi_i)\|^2 < \infty$$

for some (hence every) orthonormal basis $\{\phi_i\}$ of H.

(3) The set of all Hilbert-Schmidt operators is a two-sided selfadjoint ideal in the algebra $B(H)$.

PROOF. Suppose $T \in B(H)$ and that there exists an orthonormal basis $\{\phi_i\}$ such that

$$\sum_i \|T(\phi_i)\|^2 = M < \infty.$$

Let $\{\psi_i\}$ be another orthonormal basis.

Then
$$\sum_i \|T(\psi_i)\|^2 = \sum_i \sum_j |(T(\psi_i), \phi_j)|^2$$

$$= \sum_i \sum_j |(\psi_i, T^*(\phi_j))|^2$$

$$= \sum_j \|T^*(\phi_j)\|^2$$

$$= \sum_j \sum_i |(T^*(\phi_j), \phi_i)|^2$$

$$= \sum_j \sum_i |(\phi_j, T(\phi_i))|^2$$

$$= \sum_i \|T(\phi_i)\|^2,$$

which completes the proof of part 1.

Next, suppose T is a Hilbert-Schmidt operator. We wish to show that

$$\sum_i \|T(\phi_i)\|^2 < \infty$$

for some orthonormal basis $\{\phi_i\}$ of H. Since T is a linear combination of 4 positive Hilbert-Schmidt operators, and since

$$\|\sum_{i=1}^4 T_i(\phi)\|^2 \le 16 \sum_{i=1}^4 \|T_i(\phi)\|^2,$$

it will suffice to show the desired inequality under the assumption that T itself is a positive operator. Thus, let $\{\lambda_n\}$ be the spectrum of T, and recall that the nonzero λ_n's are the eigenvalues for T. Since T has a purely atomic spectrum, there exists an orthonormal basis $\{\phi_i\}$ for H consisting of eigenvectors for T. Then,

$$\sum_i \|T(\phi_i)\|^2 = \sum_n \lambda_n^2 m(\lambda_n) < \infty.$$

Conversely, let T be in $B(H)$ and suppose there exists an orthonormal basis $\{\phi_i\}$ such that the inequality in part 2 holds for $T = T_1 + iT_2$. It follows from part 1 that the same inequality holds as well for $T^* = T_1 - iT_2$. It then follows that the inequality holds for the real and imaginary parts $T_1 = (T + T^*)/2$ and $T_2 = (T - T^*)/2i$ of T. It will suffice then to assume that T is selfadjoint, and we write $T = T_+ - T_-$ in terms of its positive and negative parts. Now, from part b of Exercise 11.1, it follows that the inequality in part 2 must hold for T_+ and T_-, so that it will suffice in fact to assume that T is positive. We show first that T is a compact operator. Thus, let $\{v_k\}$ be a sequence of vectors in H that converges weakly to 0, and write

$$v_k = \sum_i a_{ki}\phi_i.$$

Note that for each i, we have $0 = \lim_k(v_k, \phi_i) = \lim_k a_{ki}$. Let M be an upper bound for the sequence $\{\|v_k\|\}$. Then, given $\epsilon > 0$, there exists an N such that $\sum_{i=N}^{\infty} \|T(\phi_i)\|^2 < (\epsilon/2M)^2$. Then, there exists a K such that $|a_{ki}| \leq \epsilon/2N\|T\|$ for all $1 \leq i \leq N - 1$ and all $k \geq K$. Then,

$$\|T(v_k)\| = \|T(\sum_i a_{ki}\phi_i)\|$$

$$= \|\sum_i a_{ki}T(\phi_i)\|$$

$$\leq \sum_{i=1}^{N-1} |a_{ki}|\|T(\phi_i)\| + \sum_{i=N}^{\infty} |a_{ki}|\|T(\phi_i)\|$$

$$< \epsilon/2 + \sqrt{\sum_{i=N}^{\infty} |a_{ki}|^2} \times \sqrt{\sum_{i=N}^{\infty} \|T(\phi_i)\|^2}$$

$$< \epsilon/2 + \|v_k\| \times \epsilon/2M$$

$$\leq \epsilon,$$

showing that the sequence $\{T(v_k)\}$ converges to 0 in norm. Hence, T is a (positive) compact operator. Now, using part 1 and an orthonormal basis of eigenvectors for T, we have that

$$\sum_i \lambda_i^2 m(\lambda_i) < \infty,$$

whence T is a Hilbert-Schmidt operator. This completes the proof of part 2.

We leave the verification of part 3 to an exercise.

EXERCISE 11.10. (The Space of All Hilbert-Schmidt Operators)

(a) If T is a Hilbert-Schmidt operator and S is an arbitrary element of $B(H)$, show that TS is a Hilbert-Schmidt operator. HINT: Write $S(\phi_n) = \sum_j a_{nj}\phi_j$.

(b) Show that T is a Hilbert-Schmidt operator if and only if $|T|$ is a Hilbert-Schmidt operator.

(c) Prove part 3 of the preceding theorem.

d) For T and S Hilbert-Schmidt operators, show that

$$\sum_i (T(\phi_i), S(\phi_i)) = \sum_i (S^*T(\phi_i), \phi_i)$$

exists and is independent of which orthonormal basis $\{\phi_i\}$ is used.

(e) Let $B_{hs}(H)$ denote the complex vector space of all Hilbert-Schmidt operators on H, and on $B_{hs}(H) \times B_{hs}(H)$ define

$$(T, S) = \sum_i (S^*T(\phi_i), \phi_i),$$

where $\{\phi_i\}$ is an orthonormal basis. Verify that (T, S) is a well-defined inner product on $B_{hs}(H)$, and that $B_{hs}(H)$ is a Hilbert space with respect to this inner product. This inner product is called the *Hilbert-Schmidt inner product.*

(f) If T is a Hilbert-Schmidt operator, define the *Hilbert-Schmidt norm* $\|T\|_{hs}$ of T by

$$\|T\|_{hs} = \sqrt{(T, T)} = \sqrt{\sum \|T(\phi_i)\|^2}.$$

Prove that $\|T\| \leq \|T\|_{hs}$. Show further that, if T is a Hilbert-Schmidt operator and S is an arbitrary element of $B(H)$, then

$$\|ST\|_{hs} \leq \|S\|\|T\|_{hs}.$$

(g) Show that $B_{hs}(H)$ is a Banach *-algebra with respect to the Hilbert-Schmidt norm.

THEOREM 11.7. (The Space of Trace Class Operators)

(1) *An operator $T \in B(H)$ is a trace class operator if and only if*

$$\sum_i |(T(\psi_i), \phi_i)| < \infty$$

for every pair of orthonormal sets $\{\psi_i\}$ and $\{\phi_i\}$.

(2) The set of all trace class operators is a two-sided selfadjoint ideal in the algebra $B(H)$.

(3) An operator T is a trace class operator if and only if there exist two Hilbert-Schmidt operators S_1 and S_2 such that $T = S_1 \circ S_2$.

PROOF. Since every trace class operator is a linear combination of four positive trace class operators, it will suffice, for the "only if" part of 1, to assume that T is positive. Thus, let $\{\eta_n\}$ be an orthonormal basis of eigenvectors for T, and write

$$M = \sum_n (T(\eta_n), \eta_n) = \sum_i \lambda_i m(\lambda_i),$$

where the λ_i's are the eigenvalues for T. If $\{\psi_i\}$ and $\{\phi_i\}$ are any orthonormal sets, write

$$\psi_i = \sum_n a_{ni}\eta_n,$$

where $a_{ni} = (\psi_i, \eta_n)$, and

$$\phi_i = \sum_n b_{ni}\eta_n,$$

where $b_{ni} = (\phi_i, \eta_n)$. Then

$$\sum_i |(T(\psi_i), \phi_i)| = \sum_i |\sum_n \sum_m a_{ni}\overline{b_{mi}}(T(\eta_n), \eta_m)|$$

$$= \sum_i |\sum_n a_{ni}\overline{b_{ni}}(T(\eta_n), \eta_n)|$$

$$\leq \sum_n (T(\eta_n), \eta_n) \times \sqrt{\sum_i |a_{ni}|^2} \times \sqrt{\sum_i |b_{ni}|^2}$$

$$= \sum_n (T(\eta_n), \eta_n) \times \sqrt{\sum_i |(\eta_n, \psi_i)|^2} \times \sqrt{\sum_i |(\eta_n, \phi_i)|^2}$$

$$\leq \sum_n (T(\eta_n), \eta_n)\|\eta_n\|\|\eta_n\|$$

$$= M,$$

showing that the condition in 1 holds. We leave the converse to the exercises.

It clearly follows from part 1 that the set of trace class operators forms a vector space, and it is equally clear that if T is a trace class operator, i.e., satisfies the inequality in 1, then T^* is also a trace class operator. To see that the trace class operators form a two-sided selfadjoint ideal, it will suffice then to show that ST is a trace class operator whenever $S \in B(H)$ and T is a positive trace class operator. Thus, let $\{\eta_n\}$ be an orthonormal basis of eigenvectors for T, and let $\{\psi_i\}$ and $\{\phi_i\}$ be arbitrary orthonormal sets. Write

$$\psi_i = \sum_n a_{ni}\eta_n$$

and

$$S^*(\phi_i) = \sum_n b_{mi}\eta_m.$$

Then

$$\sum_i |(ST(\psi_i), \phi_i)| = \sum_i |\sum_n \sum_m a_{ni}\overline{b_{mi}}(T(\eta_n), \eta_m)|$$

$$= \sum_i |\sum_n a_{ni}\overline{b_{ni}}(T(\eta_n), \eta_n)|$$

$$\leq \sum_i \sum_n |a_{ni}b_{ni}|(T(\eta_n), \eta_n)$$

$$\leq \sum_n (T(\eta_n), \eta_n)$$

$$\times \sqrt{\sum_i |a_{ni}|^2} \sqrt{\sum_k |b_{n,k}|^2}$$

$$= \sum_n (T(\eta_n), \eta_n)$$

$$\times \sqrt{\sum_i |(\eta_n, \psi_i)|^2} \sqrt{\sum_k |(\eta_n, S^*(\phi_k))|^2}$$

$$= \sum_n (T(\eta_n), \eta_n)\|\eta_n\|\|S(\eta_n)\|$$

$$\leq \|S\| \sum_n (T(\eta_n), \eta_n)$$

$$< \infty,$$

showing, by part 1, that ST is trace class. This completes the proof of part 2.

We leave the proof of part 3 to the following exercise.

EXERCISE 11.11. (Completing the Preceding Proof)
(a) Suppose T is a positive operator. Show that

$$\sum_j (T(\psi_j), \psi_j) = \sum_n (T(\phi_n), \phi_n)$$

for any pair of orthonormal bases $\{\psi_j\}$ and $\{\phi_n\}$. Suppose next that

$$\sum_n (T(\phi_n), \phi_n) < \infty.$$

Prove that $\sqrt{T}$ is a Hilbert-Schmidt operator, and deduce from this that T is a trace class operator.

(b) Suppose T is a selfadjoint operator, and write $T = T_+ - T_-$ in terms of its positive and negative parts. Assume that $\sum_n |(T(\phi_n), \phi_n)| < \infty$ for every orthonormal set $\{\phi_n\}$. Prove that T is a trace class operator. HINT: Choose the orthonormal set to be a basis for the closure of the range of T_+.

(c) Prove the rest of part 1 of the preceding theorem.

(d) Prove that T is a trace class operator if and only if $|T|$ is a trace class operator.

(e) Prove part 3 of the preceding theorem.

EXERCISE 11.12. (The Space of Trace Class Operators)
(a) If T is a trace class operator, define

$$\|T\|_{\mathrm{tr}} = \sup_{\{\psi_n\},\{\phi_n\}} \sum_n |(T(\psi_n), \phi_n)|,$$

where the supremum is taken over all pairs of orthonormal sets $\{\psi_n\}$ and $\{\phi_n\}$. Prove that the assignment $T \to \|T\|_{\mathrm{tr}}$ is a norm on the set $B_{\mathrm{tr}}(H)$ of all trace class operators. This norm is called the *trace class norm*.

(b) If T is a trace class operator and $\{\phi_n\}$ is an orthonormal basis, show that the infinite series $\sum(T(\phi_n), \phi_n)$ is absolutely summable. Show further that

$$\sum_n (T(\phi_n), \phi_n) = \sum_n (T(\psi_n), \psi_n),$$

where $\{\phi_n\}$ and $\{\psi_n\}$ are any two orthonormal bases. We define the *trace* $\mathrm{tr}(T)$ of a trace class operator T by

$$\mathrm{tr}(T) = \sum_n (T(\phi_n), \phi_n),$$

where $\{\phi_n\}$ is an orthonormal basis.

(c) Let T be a positive trace class operator. Show that $\|T\|_{\mathrm{tr}} = \mathrm{tr}(T)$. For an arbitrary trace class operator T, show that $\|T\|_{\mathrm{tr}} = \mathrm{tr}(|T|)$. HINT: Expand everything in terms of an orthonormal basis consisting of eigenvectors for $|T|$.

(d) Let T be a trace class operator and S be an element of $B(H)$. Prove that

$$\|ST\|_{\mathrm{tr}} \leq \|S\|\|T\|_{\mathrm{tr}}.$$

(e) Show that $B_{\mathrm{tr}}(H)$ is a Banach *-algebra with respect to the norm defined in part a.

EXERCISE 11.13. (a) Let (S, μ) be a σ-finite measure space. Show that if a nonzero multiplication operator m_f on $L^2(\mu)$ is a compact operator, then μ must have some nontrivial atomic part. That is, there must exist at least one point $x \in S$ such that $\mu(\{x\}) > 0$.

(b) Suppose μ is a purely atomic σ-finite measure on a set S. Describe the set of all functions f for which m_f is a compact operator, a Hilbert-Schmidt operator, a trace class operator, or a finite rank operator.

(c) Show that no nonzero convolution operator K_f on $L^2(\mathbb{R})$ is a compact operator. HINT: Examine the operator $U \circ K_f \circ U^{-1}$, for U the L^2 Fourier transform.

(d) Let (S, μ) be a σ-finite measure space. Suppose $k(x, y)$ is a kernel on $S \times S$, and assume that $k \in L^2(\mu \times \mu)$. Prove that the integral operator K, determined by the kernel k, is a Hilbert-Schmidt operator, whence is a compact operator.

(e) Let (S, μ) be a σ-finite measure space, and let T be a positive Hilbert-Schmidt operator on $L^2(\mu)$. Suppose $\{\phi_1, \phi_2, \dots\}$ is an orthonormal basis of $L^2(\mu)$ consisting of eigenfunctions for T, and let λ_i denote the eigenvalue corresponding to ϕ_i. Define a kernel $k(x, y)$ on $S \times S$ by

$$k(x, y) = \sum_{i=1}^{\infty} \lambda_i \phi_i(x)\overline{\phi_i(y)}.$$

Show that $k \in L^2(\mu \times \mu)$ and that T is the integral operator determined by the kernel k. Show in general that, if T is a Hilbert-Schmidt operator on $L^2(\mu)$, then there exists an element $k \in L^2(\mu \times \mu)$ such that

$$Tf(x) = \int k(x, y)f(y)\,d\mu(y)$$

for all $f \in L^2(\mu)$. Conclude that there is a linear isometry between the Hilbert space $L^2(\mu \times \mu)$ and the Hilbert space $B_{\text{hs}}(L^2(\mu))$ of all Hilbert-Schmidt operators on $L^2(\mu)$.

(f) Let S be a compact topological space, let μ be a σ-finite measure on S, and let k be a continuous function on $S \times S$. Suppose $\phi \in L^2(\mu)$ is an eigenfunction for the integral operator T determined by the kernel k. Prove that ϕ may be assumed to be continuous.

(g) (Mercer's Theorem) Let S, μ, k, and T be as in part f. Suppose T is a positive trace class operator. Prove that

$$\text{tr}(T) = \int_S k(x,x)\, d\mu(x).$$

We turn next to an examination of "unbounded selfadjoint" operators. Our definition is derived from a generalization of the properties of bounded selfadjoint operators as described in Theorem 8.7.

DEFINITION. A linear transformation T from a subspace D of a Hilbert space H into H is called an *unbounded selfadjoint operator* on H if

(1) D is a proper dense subspace of H.

(2) T is not continuous on D.

(3) T is symmetric on D; i.e., $(T(x), y) = (x, T(y))$ for all $x, y \in D$.

(4) Both $I + iT$ and $I - iT$ map D onto H.

If, in addition, $(T(x), x) \geq 0$ for all $x \in D$, then T is called an *unbounded positive operator* on H.

The subspace D is called the *domain* of T.

THEOREM 11.8. (Spectral Theorem for Unbounded Selfadjoint Operators) *Let H be a (separable and complex) Hilbert space.*

(1) *If T is an unbounded selfadjoint operator on H, then there exists a unique H-projection-valued measure p on $(\mathbb{R}, \mathcal{B})$ such that T is the integral with respect to p of the unbounded function $f(\lambda) = \lambda$; i.e., $T = \int \lambda\, dp(\lambda)$. See Theorem 9.8. Further, p is not supported on any compact interval in $\mathbb{R}$.*

(2) *If p is an H-projection-valued measure on $(\mathbb{R}, \mathcal{B})$, that is not supported on any compact interval in $\mathbb{R}$, then $T = \int \lambda\, dp(\lambda)$ is an unbounded selfadjoint operator.*

(3) *The map $p \rightarrow \int \lambda\, dp(\lambda)$ of part 2 is a 1-1 correspondence between the set of all H-projection-valued measures p on $(\mathbb{R}, \mathcal{B})$ that are not supported on any compact interval in $\mathbb{R}$ and the set of all unbounded selfadjoint operators T on H.*

PROOF. Part 2 follows from Theorem 9.8. To see part 1, let $T : D \rightarrow$ H be an unbounded selfadjoint operator, and note that $I \pm iT$ is norm-increasing on D, whence is 1-1 and onto H. Define $U = (I-iT)(I+iT)^{-1}$. Then U maps H onto itself and is an isometry. For if $y = (I + iT)^{-1}(x)$, then $x = (I + iT)(y)$, whence $\|x\|^2 = \|y\|^2 + \|T(y)\|^2$. But then

$$\begin{aligned} \|U(x)\|^2 &= \|(I - iT)(y)\|^2 \\ &= \|y\|^2 + \|T(y)\|^2 \\ &= \|x\|^2. \end{aligned}$$

Moreover,

$$I + U = (I + iT)(I + iT)^{-1} + (I - iT)(I + iT)^{-1} = 2(I + iT)^{-1},$$

showing that $I + U$ maps H 1-1 and onto D. Similarly, we see that

$$I - U = 2iT(I + iT)^{-1},$$

whence

$$T = -i(I - U)(I + U)^{-1}.$$

This unitary operator U is called the *Cayley transform* of T.

By the Spectral Theorem for normal operators, we have that $U = \int \mu \, dq(\mu)$, where q is the spectral measure for U. Because U is unitary, we know that q is supported on the unit circle T^1 in $\mathbb{C}$, and because $I + U = 2(I + iT)^{-1}$ is 1-1, we know that -1 is not an eigenvalue for U. Therefore, $q_{\{-1\}} = 0$, and the function h defined on $T^1 - \{-1\}$ by

$$h(\mu) = -i(1 - \mu)/(1 + \mu)$$

maps onto the real numbers $\mathbb{R}$. Defining $S = \int f(\mu) \, dq(\mu)$, we see from Theorem 9.8 that S is an unbounded selfadjoint operator on H. By part c of Exercise 9.15, we have that

$$\int (1/(1 + \mu)) \, dq(\mu) = (I + U)^{-1},$$

and hence that

$$S = -i(I - U)(I + U)^{-1} = T.$$

Finally, let $p = h_*(q)$ be the projection-valued measure defined on $(\mathbb{R}, \mathcal{B})$ by

$$p_E = h_*(q)_E = q_{h^{-1}(E)}.$$

From part e of Exercise 9.15, we then have that

$$\int \lambda \, dp(\lambda) = \int h(\mu) \, dq(\mu) = S = T,$$

as desired.

We leave the uniqueness of p to the exercise that follows. Part 3 is then immediate from parts 1 and 2.

DEFINITION. Let $T : D \to H$ be an unbounded selfadjoint operator and let p be the unique projection-valued measure on $(\mathbb{R}, \mathcal{B})$ for which $T = \int \lambda \, dp(\lambda)$. The projection-valued measure p is called the *spectral measure* for T.

EXERCISE 11.14. (a) Prove the uniqueness assertion in part 1 of the preceding theorem.

(b) Let $T : D \to H$ be an unbounded selfadjoint operator, let p be its spectral measure, and let $U = (I - iT)(I + iT)^{-1}$ be its Cayley transform. Prove that

$$U = \int [(1 - i\lambda)/(1 + i\lambda)] \, dp(\lambda).$$

(c) Show that there is a 1-1 correspondence between the set of all projection-valued measures on $(\mathbb{R}, \mathcal{B})$ and the set of all (bounded or unbounded) selfadjoint operators on a Hilbert space H.

DEFINITION. Let T be an unbounded selfadjoint operator with domain D. A complex number λ is said to belong to the *resolvent* of T if the linear transformation $\lambda I - T$ maps D 1-1 and onto H and $(\lambda I - T)^{-1}$ is a bounded operator on H. The *spectrum* $\mathrm{sp}(T)$ of T is the complement of the resolvent of T.

If f is a real-valued (bounded or unbounded) Borel function on $\mathbb{R}$, we write $f(T)$ for the operator $\int f(\lambda) dp(\lambda)$.

As in the case of a bounded normal operator, we make analogous definitions of *point spectrum, continuous spectrum, discrete spectrum,* and *essential spectrum.*

The following exercise is the natural generalization of Exercise 11.4 and Theorem 11.4 to unbounded selfadjoint operators.

EXERCISE 11.15. Let T be an unbounded selfadjoint operator. Verify the following:

(a) The spectral measure p for T is supported on the spectrum of T; the spectrum of T is contained in the set of real numbers; if E is a closed subset of $\mathbb{C}$ for which $p_E = I$, then E contains the spectrum of T.

(b) $\lambda \in \mathrm{sp}(T)$ if and only if there exists a sequence $\{v_n\}$ of unit vectors in H such that
$$\lim \|T(v_n) - \lambda v_n\| = 0.$$

(c) $\lambda \in \mathrm{sp}_p(T)$ if and only if λ is an eigenvalue for T, i.e., if and only if there exists a nonzero vector $v \in D$ such that $T(v) = \lambda v$.

(d) $\lambda \in \mathrm{sp}_e(T)$ if and only if there exists a sequence $\{v_n\}$ of orthonormal vectors for which

$$\lim \|T(v_n) - \lambda v_n\| = 0.$$

(e) T is an unbounded positive operator if and only if $\mathrm{sp}(T)$ is a subset of the set of nonnegative real numbers.

THEOREM 11.9. (Invariance of the Essential Spectrum under a Compact Perturbation) *Let $T : D \to H$ be an unbounded selfadjoint operator on a Hilbert space H, and let K be a compact selfadjoint operator on H. Define $T' : D \to H$ by $T' = T + K$. Then T' is an unbounded selfadjoint operator, and*

$$\mathrm{sp}_e(T') = \mathrm{sp}_e(T).$$

That is, the essential spectrum is invariant under "compact perturbations."

EXERCISE 11.16. Prove Theorem 11.9.

EXERCISE 11.17. (a) Let T be an unbounded selfadjoint operator with domain D on a Hilbert space H. Prove that the graph of T is a closed subset of $H \times H$.

(b) Let $H = L^2([0,1])$, let D be the subspace of H consisting of the absolutely continuous functions f, whose derivative f' belongs to H and for which $f(0) = f(1)$. Define $T : D \to H$ by $T(f) = if'$. Prove that T is an unbounded selfadjoint operator on H. HINT: To show that $I \pm iT$ is onto, you must find a solution to the first order linear differential equation:
$$y' \pm y = f.$$

(c) Let $H = L^2([0,1])$, let D be the subspace of H consisting of the absolutely continuous functions f, whose derivative f' belongs to H and for which $f(0) = f(1) = 0$. Define $T : D \to H$ by $T(f) = if'$. Prove that T is not an unbounded selfadjoint operator.

(d) Let $H = L^2([0,1])$, let D be the subspace of H consisting of the absolutely continuous functions f, whose derivative f' belongs to H and for which $f(0) = 0$. Define $T : D \to H$ by $T(f) = if'$. Prove that T is not an unbounded selfadjoint operator.

We give next a different characterization of unbounded selfadjoint operators. This characterization essentially deals with the size of the domain D of the operator and is frequently given as the basic definition of an unbounded selfadjoint operator. This characterization is also a useful means of determining whether or not a given $T : D \to H$ is an unbounded selfadjoint operator.

THEOREM 11.10. *Let D be a dense subspace of a separable Hilbert space H, and let $T : D \to H$ be a symmetric linear transformation of D into H. Then T is an unbounded selfadjoint operator if and only if the following condition on the domain D holds: If $x \in H$ is such that the function $y \to (T(y), x)$ is continuous on D, then x belongs to D.*

PROOF. Suppose $T : D \to H$ is an unbounded selfadjoint operator and that an $x \in H$ satisfies the given condition. Then the map sending $y \in D$ to $((I + iT)(y), x)$ is continuous on D, and so has a unique continuous extension to all of H. By the Riesz Representation Theorem for Hilbert spaces, there exists a $w \in H$ such that

$$((I + iT)(y), x) = (y, w)$$

for all $y \in D$. Since $I - iT$ maps D onto H, there exists a $v \in D$ such that $w = (I - iT)(v)$. Therefore,

$$\begin{aligned} ((I + iT)(y), x) &= (y, w) \\ &= (y, (I - iT)(v)) \\ &= ((I + iT)(y), v) \end{aligned}$$

for all $y \in D$, showing that $(z, x) = (z, v)$ for all $z \in H$, whence $x = v$, and $x \in D$.

Conversely, assume that the condition holds. We must show that T is an unbounded selfadjoint operator. We must verify that $I \pm iT$ maps D onto H. We show first that the range of $I + iT$ is dense. Thus, let x be a vector orthogonal to the range of $I + iT$. Then the map $y \to ((I+iT)(y), x)$ is identically 0 on D, showing that $(T(y), x) = i(y, x)$, and therefore the map $y \to (T(y), x)$ is continuous on D. By the condition, $x \in D$, and we have

$$0 = ((I + iT)(x), x) = (x, x) + i(T(x), x),$$

implying that $\|x\|^2 = -i(T(x), x)$, which implies that $x = 0$ since $(T(x), x)$ is real. Hence, the range of $I + iT$ is dense in H. Of course, a similar argument shows that the range of $I - iT$ is dense in H.

To see that the range of $I + iT$ is closed, let $y \in H$, and suppose $y = \lim y_n$, where each $y_n = (I + iT)(x_n)$ for some $x_n \in D$. Now the sequence $\{y_n\}$ is a Cauchy sequence, and, since $I + iT$ is norm-increasing, it follows that the sequence $\{x_n\}$ also is a Cauchy sequence. Let $x = \lim x_n$. Then, for any $z \in D$, we have

$$
\begin{aligned}
(T(z), x) &= \lim(T(z), x_n) \\
&= \lim(z, T(x_n)) \\
&= \lim(z, (1/i)(y_n - x_n)) \\
&= (z, (1/i)(y - x)),
\end{aligned}
$$

which shows that the map $z \to (T(z), x)$ is a continuous function of z. Therefore, $x \in D$, and

$$
(z, T(x)) = (T(z), x) = (z, (1/i)(y - x)),
$$

showing that $T(x) = (1/i)(y - x)$, or $(I + iT)(x) = y$, and y belongs to the range of $I + iT$. Again, a similar argument shows that the range of $I - iT$ is closed, and therefore T is an unbounded selfadjoint operator.

REMARK. We see from the preceding exercise and theorem that a symmetric operator $T : D \to H$ can fail to be an unbounded selfadjoint operator simply because its domain is not quite right. The following exercise sheds some light on this observation and leads us to the notion of "essentially selfadjoint" operators.

EXERCISE 11.18. Let H be a separable Hilbert space, and let $T : D \to H$ be a symmetric linear transformation from a dense subspace D of H into H.

(a) Suppose D' is a proper subspace of D. Show that $T : D' \to H$ can never be an unbounded selfadjoint operator. (No smaller domain will do.)

(b) Let G denote the graph of T, thought of as a subset of $H \times H$. Prove that the closure $\bar{G}$ of G is the graph of a linear transformation $S : D'' \to H$. Show further that $D \subseteq D''$, that S is an extension of T, and that S is symmetric on D''. This linear transformation S is called the *closure* of T and is denoted by $\bar{T}$. T is called *essentially selfadjoint* if $\bar{T}$ is selfadjoint.

(c) Suppose $D \subseteq E$ and that $V : E \to H$ is an unbounded selfadjoint operator. We say that V is a *selfadjoint extension* of T if V is an extension of T. Prove that any selfadjoint extension of T is an extension of $\bar{T}$. That is, $\bar{T}$ is the minimal selfadjoint extension of T.

(d) Determine whether or not the operators in parts c and d of Exercise 11.17 have selfadjoint extensions and/or are essentially selfadjoint.

EXERCISE 11.19. Let H be a separable Hilbert space.

(a) (Stone's Theorem) Let $t \to U_t$ be a homomorphism of the additive group $\mathbb{R}$ into the group of unitary operators on H. Assume that for each pair of vectors $x, y \in H$ the function $t \to (U_t(x), y)$ is continuous. Prove that there exists a unique unbounded selfadjoint operator A on H such that

$$U_t = e^{itA}$$

for all $t \in \mathbb{R}$. The operator A is called the *generator* of the one-parameter group U_t.

(b) Let A be an unbounded positive operator on H with domain D. For each nonnegative t define

$$P_t = e^{-tA}$$

Prove that the P_t's form a continuous semigroup of contraction operators. That is, show that each P_t is a bounded operator of norm ≤ 1 and that $P_{t+s} = P_t \circ P_s$ for all $t, s \geq 0$. Further, show that

$$A(x) = \lim_{t \to 0} \frac{P_t(x) - x}{t}$$

for every $x \in D$.

We conclude this chapter by summarizing our progress toward finding a mathematical model for experimental science. No proofs will be supplied for the theorems we quote here, and we emphasize that this is only a brief outline.

We have seen in Chapter VIII that the set $\mathcal{P}$ of all projections on an infinite-dimensional complex Hilbert space H could serve as a model for the set Q of all questions. Of course, many other sets also could serve as a model for Q, but we use this set $\mathcal{P}$.

Each observable A is identified with a question-valued measure, so in our model the observables are represented by projection-valued measures on $\mathbb{R}$, and we have just seen that these projection-valued measures are

in 1-1 correspondence with all (bounded and unbounded) selfadjoint operators. So, in our model, the observables are represented by selfadjoint
operators.

What about the states? How are they represented in this model? In
Chapter VII we have seen that each state α determines a character μ_α of
the set Q of questions. To see how states are represented in our model,
we must then determine what the characters of the set $\mathcal{P}$ are.

THEOREM 11.11. (Gleason's Theorem) *Let H be a separable in-*
finite dimensional complex Hilbert space, and let $\mathcal{P}$ denote the set of all
projections on H. Suppose μ is a mapping of $\mathcal{P}$ into $[0,1]$ that satisfies:

(1) *If $p \le q$, then $\mu(p) \le \mu(q)$.*
(2) *$\mu(I - p) = 1 - \mu(p)$ for every $p \in \mathcal{P}$.*
(3) *If $\{p_i\}$ is a pairwise orthogonal (summable) sequence of projec-*
tions, then $\mu(\sum p_i) = \sum \mu(p_i)$.

Then there exists a positive trace class operator S on H, for which
$\|S\|_{\mathrm{tr}} = \mathrm{tr}(S) = 1$, such that $\mu(p) = \mathrm{tr}(Sp)$ for every $p \in \mathcal{P}$.

Hence, the states are represented by certain positive trace class operators. Since each such positive trace class operator S with $\mathrm{tr}(S) = 1$ is
representable in the form

$$S = \sum \lambda_i p_i,$$

where $\sum \lambda_i M(\lambda_i) = 1$, we see that the pure states correspond to operators that are in fact projections onto 1-dimensional subspaces. Let α be
a pure state, and suppose it corresponds to the projection q_v onto the
1-dimensional subspace spanned by the unit vector v. Let A be an observable (unbounded selfadjoint operator), and suppose A corresponds
to the projection-valued measure $E \to p_E$. That is $A = \int \lambda \, dp(\lambda)$. Then
we have

$$
\begin{aligned}
\mu_{\alpha,A}(E) &= \mu_{\alpha,\chi_E(A)}(\{1\}) \\
&= \mu_\alpha(\chi_E(A)) \\
&= \mu_\alpha(q_E^A) \\
&= \mu_\alpha(p_E) \\
&= \mathrm{tr}(q_v p_E) \\
&= (q_v p_E(v), v) \\
&= (p_E(v), v) \\
&= \mu_v(E).
\end{aligned}
$$

If we regard the probability measure μ_v as being the probability distribution corresponding to a random variable X, then

$$(A(v), v) = ([\int \lambda \, dp(\lambda)](v), v) = \int \lambda \, d\mu_v(\lambda) = E[X],$$

where $E[X]$ denotes the expected value of the random variable X. We may say then that in our model $(A(v), v)$ represents the expected value of the observable A when the system is in the pure state corresponding to the projection onto the 1-dimensional subspace spanned by v.

How are time evolution and symmetries represented in our model? We have seen that these correspond to automorphisms ϕ'_t and π'_g of the set Q. So, we must determine the automorphisms of the set $\mathcal{P}$ of projections.

THEOREM 11.12. (Wigner's Theorem) *Let H be a separable infinite dimensional complex Hilbert space, and let $\mathcal{P}$ denote the set of all projections on H. Suppose η is a mapping of $\mathcal{P}$ into itself that satisfies:*

(1) *If $p \leq q$, then $\eta(p) \leq \eta(q)$.*
(2) *$\eta(I - p) = I - \eta(p)$ for every $p \in \mathcal{P}$.*
(3) *If $\{p_i\}$ is a pairwise orthogonal (summable) sequence of projections, then $\{\eta(p_i)\}$ is a pairwise orthogonal sequence of projections, and*
$$\eta(\sum p_i) = \sum \eta(p_i).$$

Then there exists a real-linear isometry U of H onto itself such that $\eta(p) = UpU^{-1}$ for all $p \in \mathcal{P}$. Further, U either is complex linear or it is conjugate linear.

Applying Wigner's Theorem to the automorphisms ϕ'_t, it follows that there exists a map $t \to U_t$ from the set of nonnegative reals into the set of real-linear isometries on H such that $\phi'_t(p) = U_t p U_t^{-1}$ for every $p \in \mathcal{P}$. Also, if G denotes a group of symmetries, then there exists a map $g \to V_g$ of G into the set of real-linear isometries of H such that $\pi'_g(p) = V_g p V_g^{-1}$ for every $p \in \mathcal{P}$.

THEOREM 11.13.

(1) *The transformations U_t can be chosen to be (complex linear) unitary operators that satisfy*

$$U_{t+s} = U_t \circ U_s$$

for all $t, s \geq 0$.

(2) *The transformations V_G can be chosen to satisfy*

$$V_{g_1 g_2} = \sigma(g_1, g_2) V_{g_1} \circ V_{g_2}$$

for all $g_1, g_2 \in G$, where $\sigma(g_1, g_2)$ is a complex number of absolute value 1. Such a map $g \to V_g$ is called a representation of G.

(3) *The operators U_t commute with the operators V_g; i.e.,*

$$U_t \circ V_g = V_g \circ U_t$$

for all $g \in G$ and all $t \geq 0$.

We have thus identified what mathematical objects will represent the elements of our experimental science, but much remains to specify. Depending on the system and its symmetries, more precise descriptions of these objects are possible. One approach is the following:

(1) Determine what the group G of all symmetries is.
(2) Study what kinds of mappings $g \to V_g$, satisfying the conditions in the preceding theorem, there are. Perhaps there are only a few possibilities.
(3) Fix a particular representation $g \to V_g$ of G and examine what operators commute with all the V_g's. Perhaps this is a small set.
(4) Try to determine, from part 3, what the transformations U_t should be.

Once the evolution transformations ϕ_t' are specifically represented by unitary operators U_t, we will be in a good position to make predictions, which is the desired use of our model Indeed, if α is a state of the system, and if α is represented in our model by a trace class operator S, then the state of the system t units of time later will be the one that is represented by the operator $U_t^{-1} S U_t$.

CHAPTER XII

NONLINEAR FUNCTIONAL ANALYSIS, INFINITE-DIMENSIONAL CALCULUS

DEFINITION Let E and F be (possibly infinite dimensional) real or complex Banach spaces, and let f be a map from a subset D of E into F. We say that f is *differentiable* at a point $x \in D$ if:

(1) x belongs to the interior of D; i.e., there exists an $\epsilon > 0$ such that $B_\epsilon(x) \subseteq D$.

(2) There exists a continuous linear transformation $L : E \to F$ and a function $\theta : B_\epsilon(0) \to F$ such that

$$f(x + h) - f(x) = L(h) + \theta(h), \tag{12.1}$$

for all $h \in B_\epsilon(0)$, and

$$\lim_{h \to 0} \|\theta(h)\| / \|h\| = 0. \tag{12.2}$$

The function f is said to be *differentiable on D* if it is differentiable at every point of D.

If $E = \mathbb{R}$, i.e., if f is a map from a subset D of $\mathbb{R}$ into a Banach space F, then f is said to have a *derivative* at a point $x \in D$ if $\lim_{t \to 0}[f(x + t) - f(x)]/t$ exists, in which case we write

$$f'(x) = \lim_{t \to 0} \frac{f(x + t) - f(x)}{t}. \tag{12.3}$$

If $D \subseteq E$, $D' \subseteq F$, and $f : D \to D'$, then f is called a *diffeomorphism* of D onto D' if f is a homeomorphism of D onto d' and f and f^{-1} are differentiable on D and D' respectively.

EXERCISE 12.1. (a) Suppose $f : D \to F$ is differentiable at a point $x \in D$, and write

$$f(x + h) - f(x) = L(h) + \theta(h)$$

as in Equation (12.1). Prove that $\theta(0) = 0$.

(b) Let $D \subseteq \mathbb{R}$, and suppose f is a function from D into a Banach space F. Show that f is differentiable at a point $x \in D$ if and only if f has a derivative at x. If f has a derivative at x, what is the continuous linear transformation $L : \mathbb{R} \to F$ and what is the map θ that satisfy Equation (12.1)?

THEOREM 12.1. *Suppose $f : D \to F$ is differentiable at a point x. Then both the continuous linear transformation L and the map θ of Equation (12.1) are unique.*

PROOF. Suppose, as in Equations (12.1) and (12.2), that

$$f(x + h) - f(x) = L_1(h) + \theta_1(h),$$

$$f(x + h) - f(x) = L_2(h) + \theta_2(h),$$

$$\lim_{h \to 0} \|\theta_1(h)\|/\|h\| = 0,$$

and

$$\lim_{h \to 0} \|\theta_2(h)\|/\|h\| = 0.$$

Then

$$L_1(h) - L_2(h) = \theta_2(h) - \theta_1(h).$$

If $L_1 \neq L_2$, choose a unit vector $u \in E$ such that $\|L_1(u) - L_2(u)\| = c > 0$. But then,

$$
\begin{aligned}
0 &= \lim_{t \to 0} (\|\theta_2(tu)\|/\|tu\| + \|\theta_1(tu)\|/\|tu\|) \\
&\geq \lim_{t \to 0} \|\theta_2(tu) - \theta_1(tu)\|/\|tu\| \\
&= \lim_{t \to 0} \|L_1(tu) - L_2(tu)\|/\|tu\| \\
&= \lim_{t \to 0} |t|c/(|t|\|u\|) \\
&= c \\
&> 0,
\end{aligned}
$$

which is a contradiction. Therefore, $L_1 = L_2$, whence $\theta_1 = \theta_2$ as well.

DEFINITION. Suppose $f : D \to F$ is differentiable at a point x. The (unique) continuous linear transformation L is called the *differential of f at x*, and is denoted by df_x. The differential is also called the *Frechet derivative of f at x*.

THEOREM 12.2. *Let E and F be real or complex Banach spaces.*

(1) *Let $f : E \to F$ be a constant function; i.e., $f(x) \equiv y_0$. Then f is differentiable at every $x \in E$, and df_x is the zero linear transformation for all x.*

(2) *Let f be a continuous linear transformation from E into F. Then f is differentiable at every $x \in E$, and $df_x = f$ for all $x \in E$.*

(3) *Suppose $f : D \to F$ and $g : D' \to F$ are both differentiable at a point x. Then $f + g : D \cap D' \to F$ is differentiable at x, and $d(f + g)_x = df_x + dg_x$.*

(4) *If $f : D \to F$ is differentiable at a point x, and if c is a scalar, then the function $g = cf$ is differentiable at x and $dg_x = cdf_x$.*

(5) *If $f : D \to F$ is differentiable at a point x, and if v is a vector in E, then*

$$df_x(v) = \lim_{t \to 0} \frac{f(x + tv) - f(x)}{t}.$$

(6) *Suppose f is a function from a subset $D \subseteq \mathbb{R}$ into F. If f is differentiable at a point x (equivalently, f has a derivative at x), then*

$$f'(x) = df_x(1).$$

PROOF. If $f(x) \equiv y_0$, then we have

$$f(x + h) - f(x) = 0 + 0;$$

i.e., we may take both L and θ to be 0. Both Equations (12.1) and (12.2) are satisfied, and $df_x = 0$ for every x.

If f is itself a continuous linear transformation of E into F, then

$$f(x + h) - f(x) = f(h) + 0;$$

i.e., we may take $L = f$ and $\theta = 0$. Then both Equations (12.1) and (12.2) are satisfied, whence $df_x = f$ for every x.

To prove part 3, write

$$f(x + h) - f(x) = df_x(h) + \theta_f(h)$$

and

$$g(x + h) - g(x) = dg_x(h) + \theta_g(h).$$

Then we have

$$(f + g)(x + h) - (f + g)(x) = [df_x + dg_x](h) + [\theta_f(h) + \theta_g(h)],$$

and we may set $L = df_x + dg_x$ and $\theta = \theta_f + \theta_g$. Again, Equations (12.1) and (12.2) are satisfied, and $d(f + g)_x = df_x + dg_x$.

Part 4 is immediate.

To see part 5, suppose f is differentiable at x and that v is a vector in E. Then we have

$$\begin{aligned} df_x(v) &= \lim_{t \to 0} df_x(tv)/t \\ &= \lim_{t \to 0} \frac{f(x + tv) - f(x) - \theta(tv)}{t} \\ &= \lim_{t \to 0} \frac{f(x + tv) - f(x)}{t} + \lim_{t \to 0} \frac{\theta(tv)}{t} \\ &= \lim_{t \to 0} \frac{f(x + tv) - f(x)}{t}, \end{aligned}$$

showing part 5.

Finally, if f is a map from a subset D of $\mathbb{R}$ into a Banach space F, and if f is differentiable at a point x, then we have from part 5 that

$$df_x(1) = \lim_{t \to 0} \frac{f(x + t) - f(x)}{t},$$

which proves that $f'(x) = df_x(1)$.

EXERCISE 12.2. Show that the following functions are differentiable at the indicated points, and verify that their differentials are as given below in parentheses.

(a) $f : B(H) \to B(H)$ is given by $f(T) = T^2$.
($df_T(S) = TS + ST$.)

(b) $f : B(H) \to B(H)$ is given by $f(T) = T^n$.
($df_T(S) = \sum_{j=0}^{n-1} T^j S T^{n-1-j}$.)

(c) f maps the invertible elements of $B(H)$ into themselves and is given by $f(T) = T^{-1}$.
($df_T(S) = -T^{-1}ST^{-1}$.)

(d) Let μ be a σ-finite measure, let p be an integer > 1, and let $f : L^p(\mu) \to L^1(\mu)$ be given by $f(g) = g^p$.
($df_g(h) = pg^{p-1}h$.)

(e) Suppose E, F, and G are Banach spaces, and let $f : E \times F \to G$ be continuous and bilinear.
($df_{x,y}(z, w) = f(x, w) + f(z, y)$.)

(f) Let E, F and G be Banach spaces, let D be a subset of E, let $f : D \to F$, let $g : D \to G$, and assume that f and g are differentiable at a point $x \in D$. Define $h : D \to F \oplus G$ by $h(y) = (f(y), g(y))$. Show that h is differentiable at x.
($dh_x(v) = (df_x(v), dg_x(v))$.)

EXERCISE 12.3. Suppose D is a subset of $\mathbb{R}^n$ and that $f : D \to \mathbb{R}^k$ is differentiable at a point $x \in D$. If we express each element of $\mathbb{R}^k$ in terms of the standard basis for $\mathbb{R}^k$, then we may write f in component form as $\{f_1, \ldots, f_k\}$.

(a) Prove that each component function f_i of f is differentiable at x.

(b) If we express the linear transformation df_x as a matrix $J(x)$ with respect to the standard bases in $\mathbb{R}^N$ and $\mathbb{R}^k$, show that the ijth entry of $J(x)$ is the partial derivative of f_i with respect to the jth variable x_j evaluated at x. That is, show that

$$J(x)_{ij} = \frac{\partial f_i}{\partial x_j}(x).$$

The matrix $J(x)$ is called the *Jacobian* of f at x.

EXERCISE 12.4. Let A be a Banach algebra with identity I, and define $f : A \to A$ by $f(x) = e^x$.

(a) Prove that f is differentiable at 0, and compute df_0.

(b) Prove that f is differentiable at every $x \in A$, and compute $df_x(y)$ for arbitrary x and y.

THEOREM 12.3. *If $f : D \to F$ is differentiable at a point x, then f is continuous at x.*

PROOF. Suppose $\epsilon > 0$ is such that $B_\epsilon(x) \subseteq D$, and let y satisfy $\|y - x\| < \epsilon$. Then

$$\|f(y) - f(x)\| = \|f(x + (y - x)) - f(x)\|$$
$$= \|df_x(y - x) + \theta(y - x)\|$$
$$\leq \|df_x\|\|y - x\| + \|y - x\|\|\theta(y - x)\|/\|y - x\|,$$

which tends to 0 as y tends to x. This shows the continuity of f at x.

THEOREM 12.4. (Chain Rule) *Let E, F, and G be Banach spaces and let $D \subseteq E$ and $D' \subseteq F$. Suppose $f : D \to F$, that $g : D' \to G$, that f is differentiable at a point $x \in D$, and that g is differentiable at the point $f(x) \in D'$. Then the composition $g \circ f$ is differentiable at x, and*

$$d(g \circ f)_x = dg_{f(x)} \circ df_x.$$

PROOF. Write y for the point $f(x) \in D'$, and define the functions θ_f and θ_g by

$$f(x + h) - f(x) = df_x(h) + \theta_f(h), \tag{12.4}$$

and

$$g(y + k) - g(y) = dg_y(k) + \theta_g(k). \tag{12.5}$$

Let $\epsilon > 0$ be such that $B_\epsilon(y) \subseteq D'$, and let $\delta > 0$ be such that $B_\delta(x) \subseteq D$, that $f(B_\delta(x)) \subseteq B_\epsilon(y)$, and that

$$\|\theta_f(h)\|/\|h\| \le 1 \tag{12.6}$$

if $\|h\| < \delta$. For $\|h\| < \delta$, define $k(h) = f(x+h) - f(x)$, and observe from Equations (12.4) and (12.6) that $\|k(h)\| \le M\|h\|$, where $M = \|df_x\| + 1$.

To prove the chain rule, we must show that

$$\lim_{h \to 0} \frac{\|g(f(x + h)) - g(f(x)) - dg_{f(x)}(df_x(h))\|}{\|h\|} = 0.$$

But

$$
\begin{aligned}
& g(f(x + h)) - g(f(x)) - dg_{f(x)}(df_x(h)) \\
& = g(y + k(h)) - g(y) - dg_y(df_x(h)) \\
& = dg_y(k(h)) + \theta_g(k(h)) - dg_y(df_x(h)) \\
& = dg_y(f(x + h) - f(x)) - dg_y(df_x(h)) \\
& \quad + \theta_g(k(h)) \\
& = dg_y(\theta_f(h)) + (\theta_g(k(h))),
\end{aligned}
$$

so

$$\|g(f(x + h)) - g(f(x)) - dg_{f(x)}(df_x(h))\| \le \|dg_y\|\|\theta_f(h)\| + \|\theta_g(k(h))\|,$$

so that it will suffice to show that

$$\lim_{h \to 0} \|\theta_g(k(h))\|/\|h\| = 0.$$

If $k(h) = 0$, then $\|\theta_g(k(h))\|/\|h\| = 0$. Otherwise,

$$\frac{\|\theta_g(k(h))\|}{\|h\|} = \frac{\|k(h)\|}{\|h\|}\frac{\|\theta_g(k(h))\|}{\|k(h)\|}$$
$$\leq M\frac{\|\theta_g(k(h))\|}{\|k(h)\|},$$

so we need only show that

$$\lim_{h \to 0}\frac{\|\theta_g(k(h))\|}{\|k(h)\|} = 0.$$

But, since f is continuous at x, we have that $k(h)$ approaches 0 as h approaches 0, so that the desired result follows from Equation (12.5).

EXERCISE 12.5. Let E, F, and G be Banach spaces, and let D be a subset of E.

(a) Let $f : D \to F$ and $g : D \to G$, and suppose B is a continuous bilinear map of $F \times G$ into a Banach space H. Define $p : D \to H$ by $p(y) = B(f(y), g(y))$. Assume that f and g are both differentiable at a point $x \in D$. Show that p is differentiable at x and compute $dp_x(y)$.

(b) Derive the "Product Formula" for differentials. That is, let A be a Banach algebra, let $f : D \to A$ and $g : D \to A$, and suppose both f and g are differentiable at a point $x \in D$. Show that the product function $f(y)g(y)$ is differentiable at x, and derive the formula for its differential.

(c) Suppose E is a Hilbert space and that $f : E \to \mathbb{R}$ is defined by $f(x) = \|x\|$. Prove that f is differentiable at every nonzero x.

(d) Let $E = L^1(\mathbb{R})$, and define $f : E \to \mathbb{R}$ by $f(x) = \|x\|_1$. Show that f is not differentiable at any point.

THEOREM 12.5. (First Derivative Test) *Let E be a Banach space, let D be a subset of E, and suppose $f : D \to \mathbb{R}$ is differentiable at a point $x \in D$. Assume that the point $f(x)$ is an extreme point of the set $f(D)$. Then df_x is the 0 linear transformation. That is, if a function achieves an extreme value at a point where it is differentiable, then the differential at that point must be 0.*

PROOF. Let v be a vector in E. Since x belongs to the interior of D, we let $\epsilon > 0$ be such that $x + tv \in D$ if $|t| < \epsilon$, and define a function $h : (-\epsilon, \epsilon) \to \mathbb{R}$ by $h(t) = f(x + tv)$. Then, by the chain rule, h is differentiable at 0. Furthermore, since $f(x)$ is an extreme point of the set $f(D)$, it follows that h attains either a local maximum or a local minimum at 0. From the first derivative test in elementary calculus, we then have that $h'(0) = dh_0(1) = 0$, implying that $df_x(v) = 0$. Since this is true for arbitrary elements $v \in E$, we see that $df_x = 0$.

THEOREM 12.6. (Mean Value Theorem) *Suppose E and F are Banach spaces, D is a subset of E, and $f : D \to F$. Suppose x and y are elements of D and that the closed line segment joining x and y is contained in D. Assume that f is continuous at each point of the closed line segment joining x to y, i.e., at each point $(1-t)x + ty$ for $0 \le t \le 1$, and assume that f is differentiable at each point on the open segment joining x and y, i.e., at each point $(1-t)x + ty$ for $0 < t < 1$. Then:*

(1) *There exists a $t^* \in (0,1)$ such that*

$$\|f(y) - f(x)\| \le \|df_z(y - x)\| \le \|df_z\|\|y - x\|,$$

for $z = (1 - t^*)x + t^*y$.

(2) *If $F = \mathbb{R}$, then there exists a t^* in (0,1) such that*

$$f(y) - f(x) = df_z(y - x)$$

for $z = (1 - t^*)x + t^*y$.

PROOF. Using the Hahn-Banach Theorem, choose ϕ in the conjugate space F^* of F so that $\|\phi\| = 1$ and

$$\|f(y) - f(x)\| = \phi(f(y) - f(x)).$$

Let h be the map of $[0,1]$ into E defined by $h(t) = (1 - t)x + ty$, and observe that

$$\|f(y) - f(x)\| = \phi(f(h(1))) - \phi(f(h(0))).$$

Defining $j = \phi \circ f \circ h$, we have from the chain rule that j is continuous on $[0,1]$ and differentiable on $(0,1)$. Then, using the Mean Value Theorem from elementary calculus, we have:

$$
\begin{aligned}
\|f(y) - f(x)\| &= j(1) - j(0) \\
&= j'(t^*) \\
&= dj_{t^*}(1) \\
&= d(\phi \circ f \circ h)_{t^*}(1) \\
&= d\phi_{f(h(t^*))}(df_{h(t^*)}(dh_{t^*}(1))) \\
&= \phi(df_{h(t^*)}(dh_{t^*}(1))) \\
&= \phi(df_{h(t^*)}(y - x)),
\end{aligned}
$$

whence

$$\|f(y) - f(x)\| \leq \|\phi\| \|df_{h(t^*)}(y - x)\|$$
$$= \|df_z(y - x)\|,$$

as desired.

We leave the proof of part 2 to the exercises.

EXERCISE 12.6. (a) Prove part 2 of the preceding theorem.

(b) Define $f : [0, 1] \to \mathbb{R}^2$ by

$$f(x) = (x^3, x^2).$$

Show that part 1 of the Mean Value Theorem cannot be strengthened to an equality. That is, show that there is no t^* between 0 and 1 satisfying $f(1) - f(0) = df_{t^*}(1)$.

(c) Define D to be the subset of $\mathbb{R}^2$ given by $0 \leq x \leq 1, 0 \leq y \leq 1$, and define $f : D \to \mathbb{R}^2$ by

$$f(x, y) = (y \cos x, y \sin x).$$

Show that every point $f(x, 1)$ is an extreme point of the set $f(D)$ but that $df_{(x,1)} \neq 0$. Conclude that the first derivative test only works when the range space is $\mathbb{R}$.

DEFINITION. Let f be a map from a subset D of a Banach space E into a Banach space F. We say that f is *continuously differentiable* at a point x if f is differentiable at each point y in a neighborhood of x and if the map $y \to df_y$ is continuous at x. ($y \to df_y$ is a map from a neighborhood of $x \in E$ into the Banach space $L(E, F)$.)

The map f is *twice differentiable* at x if it is continuously differentiable at x and the map $y \to df_y$ is differentiable at x. The differential of this map $y \to df_y$ at the point x is denoted by $d^2 f_x$. The map f is *2 times continuously differentiable* at x if the map $y \to df_y$ is continuously differentiable at x.

The notions of n times continuously differentiable are defined by induction.

EXERCISE 12.7. (a) Let E and F be Banach spaces, let D be a subset of E, and suppose $f : D \to F$ is twice differentiable at a point $x \in D$. For each $v \in E$, show that $d^2 f_x(v)$ is an element of $L(E, F)$, whence for each pair (v, w) of elements in E, $[d^2 f_x(v)](w)$ is an element of F.

(b) Let f be as in part a. Show that $d^2 f_x$ represents a continuous bilinear map of $E \oplus E$ into F.

(c) Suppose f is a continuous linear transformation of E into F. Show that f is twice differentiable everywhere, and compute $d^2 f_x$ for any x.

(d) Suppose H is a Hilbert space, that $E = F = B(H)$ and that $f(T) = T^{-1}$. Show that f is twice differentiable at each invertible T, and compute $d^2 f_T$.

THEOREM 12.7. (Theorem on Mixed Partials) *Suppose E and F are Banach spaces, D is a subset of E, and $f : D \to F$ is twice differentiable at each point of D. Suppose further that f is 2 times continuously differentiable at a point $x \in D$. Then*

$$[d^2 f_x(v)](w) = [d^2 f_x(w)](v);$$

i.e., the bilinear map $d^2 f_x$ is symmetric.

PROOF. Let v and w be in E, and let $\phi \in F^*$. Write $\phi = U + iV$ in its real and imaginary parts. Then

$$U([d^2 f_x(v)](w))$$
$$= \lim_{t \to 0} U\left(\frac{[df_{x+tv} - df_x](w)}{t}\right)$$
$$= \lim_{t \to 0} \lim_{s \to 0} U\left(\frac{f(x + tv + sw) - f(x + tv) - f(x + sw) + f(x)}{st}\right)$$
$$= \lim_{t \to 0} \lim_{s \to 0} \frac{J_s(t) - J_s(0)}{st},$$

where $J_s(t) = U(f(x+sw+tv) - f(x+tv))$. Therefore, using the ordinary Mean Value Theorem on the real-valued function J_s, we have that

$$U([d^2 f_x(v)](w)) = \lim_{t \to 0} \lim_{s \to 0} J_s'(t^*)/s$$
$$= \lim_{t \to 0} \lim_{s \to 0} U(df_{x+sw+t^*v}(v) - df_{x+t^*v}(v))/s$$
$$= \lim_{t \to 0} \lim_{s \to 0} U([df_{x+t^*v+sw} - df_{x+t^*v}](v))/s$$
$$= \lim_{t \to 0} U([d^2 f_{x+t^*v}(w)](v))$$
$$= U([d^2 f_x(w)](v)),$$

because of the continuity of $d^2 f_y$ at $y = x$. A similar computation shows that

$$V([d^2 f_x(v)](w)) = V([d^2 f_x(w)](v)),$$

which implies that

$$\phi([d^2 f_x(v)](w)) = \phi([d^2 f_x(w)](v)).$$

This equality being valid for every $\phi \in F^*$ implies that

$$[d^2 f_x(v)](w) = [d^2 f_x(w)](v),$$

as desired.

EXERCISE 12.8. (Second Derivative Test) Let E and F be Banach spaces, let D be a subset of E, and suppose $f : D \to F$ is 2 times continuously differentiable at a point $x \in D$.

(a) Show that for each pair v, w of elements in E, the function

$$y \to [d^2 f_y(v)](w)$$

is continuous at x.

(b) Suppose $F = \mathbb{R}$, that f is 2 times continuously differentiable at x, that $df_x = 0$, and that the bilinear form $d^2 f_x$ is positive definite; i.e., $[d^2 f_x(v)](v) > 0$ for every nonzero $v \in E$. Prove that f attains a local minimum at x. That is, show that there exists an $\epsilon > 0$ such that if $\|y - x\| < \epsilon$ then $f(x) < f(y)$. HINT: Use the Mean Value Theorem twice to show that $f(y) - f(x) > 0$ for all y in a sufficiently small ball around x.

EXERCISE 12.9. Let (X, d) be a metric space. A map $\phi : X \to X$ is called a *contraction map* on X if there exists an α with $0 \le \alpha < 1$ such that

$$d(\phi(x), \phi(y)) \le \alpha d(x, y)$$

for all $x, y \in X$.

(a) If ϕ is a contraction map on (X, d), $x_0 \in X$, and $k < n$ are positive integers, show that

$$d(\phi^n(x_0), \phi^k(x_0)) \le \sum_{j=k}^{n-1} d(\phi^{j+1}(x_0), \phi^j(x_0))$$

$$\le \sum_{j=k}^{n-1} \alpha^j d(\phi(x_0), x_0)$$

$$= d(\phi(x_0), x_0) \alpha^k \frac{1 - \alpha^{n-k}}{1 - \alpha},$$

where ϕ^i denotes the composition of ϕ with itself i times.

(b) If ϕ is a contraction map on a complete metric space (X, d), and $x_0 \in X$, show that the sequence $\{\phi^n(x_0)\}$ has a limit in X.

(c) If ϕ is a contraction map on a complete metric space (X, d), and $x_0 \in X$, show that the limit y_0 of the sequence $\{\phi^n(x_0)\}$ is a fixed point of ϕ; i.e., $\phi(y_0) = y_0$.

(d) (Contraction mapping theorem) Show that a contraction map on a complete metric space (X, d) has one and only one fixed point y_0, and that $y_0 = \lim_n \phi^n(x)$ for each $x \in X$.

THEOREM 12.8. (Implicit Function Theorem) *Let E and F be Banach spaces, and equip $E \oplus F$ with the max norm. Let f be a map of an open subset O in $E \oplus F$ into F, and suppose f is continuously differentiable at a point $x = (x_1, x_2) \in O$. Assume further that the linear transformation $T : F \to F$, defined by $T(w) = df_x(0, w)$, is 1-1 and onto F. Then there exists a neighborhood U_1 of x_1 in E, a neighborhood U_2 of x_2 in F, and a unique continuous function $g : U_1 \to U_2$ such that*

(1) *The level set $f^{-1}(f(x)) \cap U$ coincides with the graph of g, where $U = U_1 \times U_2$.*

(2) *g is differentiable at x_1, and*

$$dg_{x_1}(h) = -T^{-1}(df_x(h, 0)).$$

PROOF. We will use the contraction mapping theorem. (See the previous exercise.) By the Isomorphism Theorem for continuous linear transformations on Banach spaces, we know that the inverse T^{-1} of T is an element of the Banach space $L(F, F)$. From the hypothesis of continuous differentiability at x, we may assume then that O is a sufficiently small neighborhood of x so that

$$\|df_z - df_x\| < 1/2\|T^{-1}\| \qquad (12.7)$$

if $z \in O$. Write

$$f(x + h) - f(x) = df_x(h) + \theta(h).$$

We may assume also that O is sufficiently small so that

$$\|\theta(h)\| \le \|h\|/2\|T^{-1}\| \qquad (12.8)$$

if $x + h \in O$. Now there exist neighborhoods O_1 of x_1 and O_2 of x_2 such that $O_1 \times O_2 \subseteq O$. Choose $\epsilon > 0$ such that the ball $B_\epsilon(x_2) \subseteq O_2$, and then choose $\delta > 0$ such that $B_\delta(x_1) \subseteq O_1$ and such that

$$\delta < \max(\epsilon, \epsilon/2\|T^{-1}\|\|df_x\|). \tag{12.9}$$

Set $U_1 = B_\delta(x_1)$, $U_2 = B_\epsilon(x_2)$, and $U = U_1 \times U_2$.

Let X be the set of all continuous functions from U_1 into U_2, and make X into a metric space by defining

$$d(g_1, g_2) = \sup_{v \in U_1} \|g_1(v) - g_2(v)\|.$$

Then, in fact, X is a complete metric space. (See the following exercise.)

Define a map ϕ, from X into the set of functions from U_1 into F, by

$$[\phi(g)](v) = g(v) - T^{-1}(f(v, g(v)) - f(x)).$$

Notice that each function $\phi(g)$ is continuous on U_1. Further, if $v \in U_1$, i.e., if $\|v - x_1\| < \delta$, then using inequalities (12.8) and (12.9) we have that

$$\|[\phi(g)](v) - x_2\|$$
$$= \|g(v) - x_2 - T^{-1}(f(v, g(v)) - f(x))\|$$
$$\leq \|T^{-1}\|\|T(g(v) - x_2) - f(v, g(v)) + f(x)\|$$
$$= \|T^{-1}\|$$
$$\quad \times \|df_x(0, g(v) - x_2) - df_x(v - x_1, g(v) - x_2) - \theta(v - x_1, g(v) - x_2)\|$$
$$= \|T^{-1}\|\|df_x(v - x_1, 0) - \theta(v - x_1, g(v) - x_2)\|$$
$$\leq \|T^{-1}\|\|df_x\|\delta + \|T^{-1}\|\|\theta(v - x_1, g(v) - x_2)\|$$
$$< \|T^{-1}\|\|df_x\|\delta + \|(v - x_1, g(v) - x_2)\|/2$$
$$< \|T^{-1}\|\|df_x\|\delta + \max(\|v - x_1\|, \|g(v) - x_2\|)/2$$
$$< \|T^{-1}\|\|df_x\|\delta + \epsilon/2$$
$$< \epsilon,$$

showing that $\phi(g) \in X$.

Next, for $g_1, g_2 \in X$, we have:

$$d(\phi(g_1), \phi(g_2))$$
$$= \sup_{v \in U_1} \|g_1(v) - g_2(v) - T^{-1}(f(v, g_1(v)) - f(v, g_2(v)))\|$$
$$\leq \sup_{v \in U_1} \|T^{-1}\|$$
$$\times \|T(g_1(v) - g_2(v)) - [f(v, g_1(v)) - f(v, g_2(v))]\|$$
$$= \sup_{v \in U_1} \|T^{-1}\|$$
$$\times \|[T(g_1(v)) - f(v, g_1(v))] - [T(g_2(v)) - f(v, g_2(v))]\|$$
$$\leq \sup_{v \in U_1} \|T^{-1}\|$$
$$\times \|J^v(w_1) - J^v(w_2)\|,$$

where $w_i = g_i(v)$, and where J^v is the function defined on O_2 by

$$J^v(w) = T(w) - f(v, w).$$

So, by the Mean Value Theorem and inequality (12.7), we have

$$d(\phi(g_1), \phi(g_2)) \leq \sup_{v \in U_1} \|T^{-1}\|\|d(J^v)_z(w_1 - w_2)\|$$
$$= \sup_{v \in U_1} \|T^{-1}\|\|[T - df_z](g_1(v) - g_2(v))\|$$
$$\leq \sup_{v \in U_1} \|T^{-1}\|\|df_x - df_z\|\|g_1(v) - g_2(v)\|$$
$$\leq d(g_1, g_2)/2,$$

showing that ϕ is a contraction mapping on X.

Let g be the unique fixed point of ϕ. Then, $\phi(g) = g$, whence $f(v, g(v)) = f(x)$ for all $v \in U_1$, which shows that the graph of g is contained in the level set $f^{-1}(f(x)) \cap U$. On the other hand, if $(v_0, w_0) \in U$ satisfies $f(v_0, w_0) = f(x)$, we may set $g_0(v) \equiv w_0$, and observe that $[\phi^n(g_0)](v_0) = w_0$ for all n. Therefore, the unique fixed point g of ϕ must satisfy $g(v_0) = w_0$, because $g = \lim \phi^n(g_0)$. Hence, any element (v_0, w_0) of the level set $f^{-1}(f(x)) \cap U$ belongs to the graph of g.

Finally, to see that g is differentiable at x_1 and has the prescribed differential, it will suffice to show that

$$\lim_{h \to 0} \|g(x_1 + h) - g(x_1) + T^{-1}(df_x(h, 0))\|/\|h\| = 0.$$

Now, because

$$f(x_1 + h, x_2 + (g(x_1 + h) - x_2)) - f(x_1, x_2) = 0,$$

we have that

$$0 = df_x(h, 0) + df_x(0, g(x_1 + h) - x_2) + \theta(h, g(x_1 + h) - x_2),$$

or

$$g(x_1 + h) - g(x_1) = -T^{-1}(df_x(h, 0)) - T^{-1}(\theta(h, g(x_1 + h) - g(x_1))).$$

Hence, there exists a constant M such that

$$\|g(x_1 + h) - g(x_1)\| \leq M\|h\|$$

whenever $x_1 + h \in U_1$. (How?) But then

$$\frac{\|g(x_1 + h) - g(x_1) + T^{-1}(df_x(h, 0))\|}{\|h\|}$$

$$\leq \frac{\|T^{-1}\|\|\theta(h, g(x_1 + h) - g(x_1))\|}{\|h\|}$$

$$\leq \frac{\|T^{-1}\|M\|\theta(h, g(x_1 + h) - g(x_1))\|}{\|(h, g(x_1 + h) - g(x_1))\|},$$

and this tends to 0 as h tends to 0 since g is continuous at x_1.
This completes the proof.

EXERCISE 12.10. Verify that the set X used in the preceding proof is a complete metric space with respect to the function d defined there.

THEOREM 12.9. (Inverse Function Theorem) *Let f be a mapping from an open subset O of a Banach space E into E, and assume that f is continuously differentiable at a point $x \in O$. Suppose further that the differential df_x of f at x is 1-1 from E onto E. Then there exist neighborhoods O_1 of x and O_2 of $f(x)$ such that f is a homeomorphism of O_1 onto O_2. Further, the inverse f^{-1} of the restriction of f to O_1 is differentiable at the point $f(x)$, whence*

$$d(f^{-1})_{f(x)} = (df_x)^{-1}.$$

PROOF. Define a map $J : E \times O \to E$ by $J(v, w) = v - f(w)$. Then J is continuously differentiable at the point $(f(x), x)$, and

$$dJ_{(f(x),x)}(0, y) = -df_x(y),$$

which is 1-1 from E onto E. Applying the implicit function theorem to J, there exist neighborhoods U_1 of the point $f(x)$, U_2 of the point x, and a continuous function $g : U_1 \to U_2$ whose graph coincides with the level set $J^{-1}(0) \cap (U_1 \times U_2)$. But this level set consists precisely of the pairs (v, w) in $U_1 \times U_2$ for which $v = f(w)$, while the graph of g consists precisely of the pairs (v, w) in $U_1 \times U_2$ for which $w = g(v)$. Clearly, then, g is the inverse of the restriction of f to U_2. Setting $O_1 = U_2$ and $O_2 = U_1$ gives the first part of the theorem. Also, from the implicit function theorem, $g = f^{-1}$ is differentiable at $f(x)$, and then the fact that $d(f^{-1})_{f(x)} = (df_x)^{-1}$ follows directly from the chain rule.

EXERCISE 12.11. Let H be a Hilbert space and let $E = B(H)$.

(a) Show that the exponential map $T \to e^T$ is 1-1 from a neighborhood U of 0 onto a neighborhood V of I.

(b) Let U and V be as in part a. Show that, for $T \in U$, we have e^T is a positive operator if and only if T is selfadjoint, and e^T is unitary if and only if T is *skewadjoint,* i.e., $T^* = -T$.

THEOREM 12.10. (Foliated Implicit Function Theorem) *Let E and F be Banach spaces, let O be an open subset of $E \times F$, and let $f : O \to F$ be continuously differentiable at every point $y \in O$. Suppose $x = (x_1, x_2)$ is a point in O for which the map $w \to df_x(0, w)$ is 1-1 from F onto F. Then there exist neighborhoods U_1 of x_1, U_2 of $f(x)$, U of x, and a diffeomorphism $J : U_1 \times U_2 \to U$ such that $J(U_1 \times \{z\})$ coincides with the level set $f^{-1}(z) \cap U$ for all $z \in U_2$.*

PROOF. For each $y \in O$, define $T_y : F \to F$ by $T_y(w) = df_y(0, w)$. Because T_x is an invertible element in $L(F, F)$, and because f is continuously differentiable at x, we may assume that O is small enough so that T_y is 1-1 and onto for every $y \in O$.

Define $h : O \to E \times F$ by

$$h(y) = h(y_1, y_2) = (y_1, f(y)).$$

Observe that h is continuously differentiable on O, and that

$$dh_x(v, w) = (v, df_x(v, w)),$$

whence, if $dh_x(v_1, w_1) = dh_x(v_2, w_2)$, then $v_1 = v_2$. But then $df_x(0, w_1 - w_2) = 0$, implying that $w_1 = w_2$, and therefore dh_x is 1-1 from $E \times F$ into $E \times F$. The exercise that follows this proof shows that dh_x is also onto, so we may apply the inverse function theorem to h. Thus, there exist neighborhoods O_1 of x and O_2 of $h(x)$ such that h is a homeomorphism of O_1 onto O_2. Now, there exist neighborhoods U_1 of x_1 and U_2 of $f(x)$ such that $U_1 \times U_2 \subseteq O_2$, and we define U to be the neighborhood $h^{-1}(U_1 \times U_2)$ of x. Define J to be the restriction of h^{-1} to $U_1 \times U_2$. Just as in the above argument for dh_x, we see that dh_y is 1-1 and onto if $y \in U$, whence, again by the inverse function theorem, J is differentiable at each point of its domain and is therefore a diffeomorphism of $U_1 \times U_2$ onto U.

We leave the last part of the proof to the following exercise.

EXERCISE 12.12. (a) Show that the linear transformation dh_x of the preceding proof is onto.

(b) Prove the last part of Theorem 12.10; i.e., show that $J(U_1 \times \{z\})$ coincides with the level set $f^{-1}(z) \cap U$.

We close this chapter with some exercises that examine the important special case when the Banach space E is actually a (real) Hilbert space.

EXERCISE 12.13. (Implicit Function Theorem in Hilbert Space) Suppose E is a Hilbert space, F is a Banach space, D is a subset of E, $f : D \to F$ is continuously differentiable on D, and that the differential df_x maps E onto F for each $x \in D$. Let c be an element of the range of f, let S denote the level set $f^{-1}(c)$, let x be in S, and write M for the kernel of df_x. Prove that there exists a neighborhood U_x of $0 \in M$, a neighborhood V_x of $x \in E$, and a continuously differentiable 1-1 function $g_x : U_x \to V_x$ such that the range of g_x coincides with the intersection $V_x \cap S$ of V_x and S. HINT: Write $E = M \oplus M^\perp$. Show also that $d(g_x)_0(h) = h$. We say that the level set $S = f^{-1}(c)$ is *locally parameterized* by an open subset of M.

DEFINITION. Suppose E is a Hilbert space, F is a Banach space, D is a subset of E, $f : D \to F$ is continuously differentiable on D, and that the differential df_x maps E onto F for each $x \in D$. Let c be an element of the range of f, and let S denote the level set $f^{-1}(c)$. We say that S is a *differentiable manifold*, and if $x \in S$, then a vector $v \in E$ is called a *tangent vector* to S at x if there exists an $\epsilon > 0$ and a continuously differentiable function $\phi : [-\epsilon, \epsilon] \to S \subseteq E$ such that $\phi(0) = x$ and $\phi'(0) = v$.

EXERCISE 12.14. Let x be a point in a differentiable manifold S, and write M for the kernel of df_x. Prove that v is a tangent vector to S at x if and only if $v \in M$. HINT: If $v \in M$, use Exercise 12.13 to define $\phi(t) = g_x(tv)$.

DEFINITION. Let D be a subset of a Banach space E, and suppose $f : D \to \mathbb{R}$ is differentiable at a point $x \in D$. We identify the conjugate space $\mathbb{R}^*$ with $\mathbb{R}$. By the *gradient* of f at x we mean the element of E^* defined by grad $f(x) = df_x^*(1)$, where df_x^* denotes the adjoint of the continuous linear transformation df_x.

If E is a Hilbert space, then grad $f(x)$ can by the Riesz representation theorem for Hilbert spaces be identified with an element of $E \equiv E^*$.

EXERCISE 12.15. Let S be a manifold in a Hilbert space E, and let g be a real-valued function that is differentiable at each point of an open set D that contains S. Suppose $x \in S$ is such that $g(x) \geq g(y)$ for all $y \in S$, and write $M = \ker(df_x)$. Prove that the vector grad $g(x)$ is orthogonal to M.

EXERCISE 12.16. (Method of Lagrange Multipliers) Let E be a Hilbert space, let D be an open subset of E, let $f = \{f_1, \ldots, f_n\}$: $D \to \mathbb{R}^n$ be continuously differentiable at each point of D, and assume that each differential df_x for $x \in D$ maps onto $\mathbb{R}^n$. Suppose g is a real-valued differentiable function on D and that g attains a maximum on S at the point x. Prove that there exist real constants $\{\lambda_1, \ldots, \lambda_n\}$ such that

$$\operatorname{grad} g(x) = \sum_{i=1}^{n} \lambda_i \operatorname{grad} f_i(x).$$

The constants $\{\lambda_i\}$ are called the *Lagrange multipliers*.

EXERCISE 12.17. Let S be the unit sphere in $L^2([0,1])$; i.e., S is the manifold consisting of the functions $f \in L^2([0,1])$ for which $\|f\|_2 = 1$.

(a) Define g on S by $g(f) = \int_0^1 f(x)\,dx$. Use the method of Lagrange multipliers to find all points where g attains its maximum value on S.

(b) Define g on S by $g(f) = \int_0^1 |f|^{3/2}(x)\,dx$. Find the maximum value of g on S.

BIBLIOGRAPHY

J. B. Conway, *A Course in Functional Analysis*, Springer-Verlag, New York, 1985.

J. M. G. Fell and R. S. Doran, *Representations of ∗-Algebras, Locally Compact Groups, and Banach ∗-Algebraic Bundles*, Academic Press, Boston, 1988.

G. B. Folland, *Harmonic Analysis in Phase Space*, Princeton University Press, Princeton, 1989.

G. B. Folland, *Real Analysis: Modern Techniques and their Applications*, John Wiley, New York, 1984.

E. Hewitt and K. A. Ross, *Abstract Harmonic Analysis*, Springer-Verlag, Berlin, 1963.

A. A. Kirillov, *Elements of the Theory of Representations*, Springer-Verlag, New York, 1976.

L. H. Loomis, *An Introduction to Abstract Harmonic Analysis*, Van Nostrand, New York, 1953.

G. W. Mackey, *The Mathematical Foundations of Quantum Mechanics*, W. A. Benjamin, New York, 1963.

H. L. Royden, *Real Analysis*, Macmillan, New York, 1988.

W. Rudin, *Functional Analysis*, McGraw-Hill, New York, 1973.

K. Yoshida, *Functional Analysis*, Springer-Verlag, Berlin, 1965.

259

INDEX

Almost everywhere p
  (a.e.p), 165
Absolute value of an
  operator, 214
Adjoining an identity, 189
Adjoint, 93
  of an operator, 155
Alaoglu's Theorem, 94
Algebraic direct sum, 2
Approximate identities, 104

Baire measures, 26
Baire sets, 26
Banach algebra, 75, 166, 175
  adjoining an identity, 189
  adjoint, 187
  Gelfand transform, 193
  $L^1(\mathbb{R})$, 188
  normal element, 187
  projection element, 187
  selfadjoint subset, 187
  structure space, 192
  structure space
    of $L^1(\mathbb{R})$, 194
  unitary element, 187
Banach *-algebra, 187

Banach limits, 38
Banach means, 38
Banach space, 65
  characterization, 66
  complex, 65
  reflexive, 93
  reflexive criterion, 96
Bessel's Inequality, 145
Basis:
  for a vector space, 2
  for a topology, 5
  orthonormal, 148
Bilateral shift, 157
Borel function, 19, 165
Borel measure, 19
Borel sets, 19, 165
Borel space, 165

C^*-algebra, 188
  sub C^*-algebra, 188
Cartesian products, 1
Cauchy-Schwarz
  Inequality, 142

Cayley transform, 157
  of an unbounded selfadjoint
    operator, 232
Chain rule, 246
Choquet's Theorem, 61
  and the Riesz Represen-
    tation Theorem, 97
Closed Graph Theorem, 71
Closure of an operator, 236
Cluster point, 35
Cofinal, 35
Compact operator, 219
  characterization, 221
  properties of the set, 221
Compact space, 5
  locally compact, 5
  σ-compact, 5, 20
Compact support, 19
Completion of a normed
    linear space, 75
Complex Banach space, 65
Complex measure, 24, 91
  norm, 91
Cone, 28
Conjugate linear, 141
Conjugate space, 90
Continuity (characterized), 49
Continuity and the graph, 50
Continuous spectrum, 215
Continuously differen-
    tiable, 249
Contraction map, 251
Contraction Mapping
    Theorem, 252
Convex hull, 54
Convex set, 54

Convolution kernels, 103, 104
  on the circle, 103
  operator, 103, 160
Convolution Theorem, 113
Countably normed space, 52
Cyclic vector, 169

Derivative, 241
Diagonalizable, 204
Diagonalizing a Hermitian
    matrix, 205
Diffeomorphism, 242
Differentiability, 241
Differentiable manifold, 257
Differential, 243
Dimension (vector space), 2
Direct product, 3
Direct sum, 3
  Hilbert spaces, 151
  normed linear spaces, 65
  projection-valued
    measures, 167
Directed set, 35
Discontinuous linear
    functionals, 50
Discrete spectrum, 215
Distribution, 86
  properties, 87
  tempered function, 86
  tempered measure, 87
Distributional derivative, 87
Distributions as derivatives
    of functions, 88
Domain of an unbounded
    selfadjoint operator, 231
Dual space, 81
Duality Theorem, 83

Eigenvalue, 156
Eigenvector, 156

Essential spectrum, 215
Essentially selfadjoint, 236
Extreme point, 59

Face, 59
Finite intersection
    property, 8
Finite rank operator, 219
Finitely additive measure, 40
    translation-invariant, 41
First derivative test, 247
Fourier transform, 112
    inversion Theorem, 114
    Gauss kernel, 113
    on $\mathbb{R}^n$, 118
    L^2 transform, 116
    Plancherel Theorem, 114
    tempered distribution, 117
Frechet derivative, 243
Frechet space, 44, 86

Gauss kernel, 108
    Fourier transform, 113
Gelfand transform, 193
General spectral
    Theorem, 200
Generator of a one-parameter
    group, 237
Gleason's Theorem, 238
Gradient, 258
Gram-Schmidt process, 145
Green's functions, 108
    for the Laplacian, 119?

Hahn-Banach Theorem:
    complex version, 41
    extreme point version, 59
    locally convex version, 56
    norm version, 33, 75

positive cone version, 28
semigroup-invariant
    version, 39
seminorm version, 32
Hausdorff space, 5
Hausdorff-Young
    Inequality, 117
Hausdorrff Maximality
    Principle, 1
Hermitian form, 141, 154
Hilbert space direct sum, 151
Hilbert transform, 120

Hilbert-Schmidt operator, 220
    characterization, 223
Hilbert-Schmidt inner
    product, 226
Hilbert-Schmidt norm, 226

Idempotent operator, 156

Implicit Function
    Theorem, 252
    in Hilbert space, 257
    foliated version, 256
Independence of spectrum, 199
Infinite Cartesian
    products, 1
Inner product, 141
Inner product space, 141
Integral kernel, 99

Integral operator, 99
    bounded, 99
Invariance of the essential
    spectrum, 234
Inverse Function
    Theorem, 255
Involution, 187
Isometric isomorphism, 65
Isomorphism Theorem, 70

Jacobian, 245

Kernel and range, 3
Kernel, 99
  convolution, 103
  Gauss, 108
  Poisson in $\mathbb{R}^n$, 107
  Poisson on the circle, 107
  Poisson on the line, 107
  reproducing, 104
  singular, 104
Krein-Milman Theorem, 60

Lagrange multipliers, 258

Linear functional, 4
  bounded 21
  conjugate linear, 141
  norm, 90
  positive, 12, 28
Linear isomorphism, 3
Linear transformation, 3
Locally compact, 5
Locally convex, 54

Matrix coefficient, 153
Max norm, 65
Mazur's Theorem, 191
Mean Value Theorem, 248
Mercer's Theorem, 231
Metric, 6
Metrizable space, 6
Minkowski functional, 56
Mixed partials, 250
Mixed state, 127
Multiplication operator, 160
Multiplicity, 215
Multipliers on the line, 120

Natural map, 4
Neighborhood, 5

Net, 35
Non-locally convex space, 58
Norm:
  of a linear functional, 90
  of an operator, 72
Normable spaces, 66
Normal operator, 156
Normal topological space, 5

Observable, 125
  compatible, 128
  simultaneously, 128
Open Mapping Theorem, 71
Orthogonal complement, 144
Orthogonal projection, 156
Orthogonality, 144
Orthonormal basis, 148
Orthonormal set, 144

Parallelogram Law, 144
Parseval's Equality, 146
Partial isometry, 213
Plancherel Theorem, 114
Point spectrum, 215

Poisson kernel:
  on the circle, 107
  on the line, 107
  in $\mathbb{R}^n$, 107
Polar decomposition, 213

Polarization Identity, 143
  for an operator, 153
Positive part of a self-
  adjoint operator, 212
Positive cone, 28
Positive functional, 12, 28
Positive operator, 156
Positive square root of a
  positive operator, 211
Product topology, 7

Projection onto a
    subspace, 150
Projection Theorem, 149
Projection-valued
    measure, 165
    canonical, 168
    change of variables, 176
    direct sum, 167
    integral of a bounded
        function, 175
    integral of an unbounded
        function, 184
    invariant subspace, 165
    $L^\infty(p)$, 166
    Riesz representation
        Theorem, 177
    unitarily equivalent, 166
Pure state, 127
Purely atomic spectrum, 215

Questions, 128
    complementary, 130
    ordering, 129
    orthogonality, 131
    sum, 130
Question-valued measure, 132
Quotient space, 4
Quotient topological vector
    space, 48
Quotient topology, 8

Reflexive Banach space, 93
    criterion, 96
Regular topological space, 5
Reproducing kernels, 104
Resolution of the
    identity, 203
Resolvent, 195, 233
Restriction of a projection-

valued measure, 165
Riemann-Lebesque
    Theorem, 113
Riesz Representation
    Theorem:
    for $C_0(\Delta)$, 21
    complex version, 25
    in Hilbert space, 151
    for projection-
        valued measures, 177
Riesz Interpolation
    Theorem, 76
Ring of sets, 40

Schwartz space $\mathcal{S}$, 53
Second countable, 5
Second derivative test, 251
Second dual, 93
Self dual, 152
Selfadjoint extension, 237
Selfadjoint operator, 156
    unbounded, 231
Selfadjoint subset, 187
Seminorm, 31
    on a complex space, 41
Separating vector, 169
Separation Theorem, 58
Sgn (signum function), 78, 117
Simultaneously diagona-
    lizable, 205
Singular kernels, 104
Spectral mapping
    Theorem, 195, 203
Spectral measure, 202
    unbounded selfadjoint
        operator, 232
Spectral radius formula, 197
Spectral radius, 195

Spectral Theorem:
  general, 200
  normal operator, 200
  selfadjoint operator, 203
  unbounded selfadjoint
    operator, 231
Spectrum, 195, 233
  characterization of point
    spectrum, 215
  continuous, 215
  discrete, 215
  essential, 215
  independence, 199
  invariance of essential
    spectrum, 234
  multiplicity, 215
  point, 215
  properties, 218
  purely atomic, 215
States, 125
  mixed, 127
  pure, 127
Stone's Axiom, 11
Stone's Theorem, 205, 237
Strong topology, 81
Structure space, 192
  of $L^1(\mathbb{R})$, 194
Subadditive functional, 31, 55
Sub C^*-algebra, 188
Subnet, 35
Supporting vector, 169
Supremum norm, 21
Symmetry, 137

Tychonoff Theorem, 36
Tangent vector, 257
Tempered distribution, 86
Tempered function, 86

Tempered measure, 87
Time evolution, 136
Topologically isomorphic, 44
Topology on a set, 5
  basis for, 5
  product, 7
  relation among the weak,
    weak*, and norm, 93
  σ-compact, 5, 20
  weak and strong, 81
  relative, 5
  strong, 81
  weak topologies and
    metrizability, 93
  weak, 7, 81
  weak*, 83
Total variation norm, 92
Trace class operator, 220
  the set, 223, 226
Trace class norm, 229
Trace of an operator, 229
Transpose, 93
Twice differentiable, 249

Urysohn's Lemma, 8
Unbounded selfadjoint
    operator, 231
  Cayley transform, 232
  domain, 231
  positive, 231
  resolvent, 233
  spectral measure, 233
  Spectral Theorem, 231
  spectrum, 233

Uniform Boundedness
    Principle, 73
Unilateral shift, 157

Unitary operator, 156
   characterization, 156

Vanish at infinity, 19
Vector lattice, 11
Vector space direct
   product, 3

Weak topology, 7, 81
Weak* topology, 93
Wigner's Theorem, 239

Young's Inequality, 102